Linda — h's
to keep

Best, Stephen M.

Praise for "Extraterrestrials"

"Such a fascinating subject!"

"…so thought provoking! "

"Stephen Mather-Lees is a true visionary of human evolution."

"A thoughtful compendium of ideas."

"The questions spark off all sorts of philosophical points."

Also by Stephen Mather-Lees:

Who The Bleep Designed Us? A God? ETs? Both? Neither?

Questing Press, 2009.

EXTRATERRESTRIALS

EXTRATERRESTRIALS

They are here. Now.

And they have been here for quite a while!

It's time to talk about them. And to them.

By

Stephen Mather-Lees

Questing Press, Atkinson, NH

Cover design by Linette Lizotte.
(linette@weichicenters.com)

Copyright © 2011 Stephen Mather-Lees

All rights reserved, including the right to reproduce in whole or in part in any form.

ISBN: 978-1-4637-0637-1

Published by:

Questing Press, Atkinson, NH 03811.   July 2011

Printed in the United States of America

This book, like "Who The Bleep Designed Us" which preceded it, is dedicated to those brave souls who challenged the religious and/or scientific establishments of their day by going public with their ideas and research results.

Some of them were vindicated in their lifetimes, some after their deaths, some never. Many risked their careers and lost them, others took the same risk and kept them.

Jesus of Nazareth is certainly the best known. Then there was Galileo, whom the Catholic Church eventually forced to recant his "heretical" idea that the Earth revolved around the Sun. Giordano Bruno was burned at the stake for the same offense. Nicola Tesla was another, so was Wilhelm Reich, and so was Eugene Mallove. There are many others too numerous to mention here – writers, teachers, inventors, and whistle blowers from all over the world. They know who they are, and we should salute them. They are responsible for a great deal of our progress as a species.

Contents

Foreword

Preface

Introduction

Chapter 1   Levels of Belief

Chapter 2   Why should we be talking to the ETs?

Chapter 3   And who are "They", anyway?

Chapter 4   How do we make contact?

Chapter 5   Will they respond?

Chapter 6   The Military Mind and Threatened Worldviews

Chapter 7   What questions should we be asking?

Chapter 8   What can we look forward to, short term?

Chapter 9   What can we look forward to, longer term?

Bibliography

Resources

x

Foreword

August T. Jaccaci

I have never seen a UFO, and to the best of my knowledge I have never been abducted by alien creatures arriving in a UFO, but I believe they are absolutely real and with us now. I have read books by the former Harvard Medical School psychiatrist Dr. John E. Mack MD which contain many accounts of patients who felt that they had been abducted, and who he helped to adjust themselves to the strangeness of their abduction experiences. I have also read Steven Jones' book "Welcome To The Dance", about his lifetime of abductions since he was five years old until nearly sixty, and his eventual fearless urging of his readers to recognize the value that aliens in UFOs bring to our lives, helping us to learn more about the precious meaning of our humanity and our truly glorious planet Earth.

Similarly I have watched a DVD film sent to me by Michael O'Connell, a Harvard trained psychologist, of a regression hypnosis session in which he guided an abduction experiencer from North Carolina. The DVD was made from a Discovery Channel TV program called "The Fayetteville Five". Michael is now a leading therapist having helped over two hundred people come to terms with their abductions.

Lest you think that this enthusiastic endorsement of Stephen Mather-Lees' book "Extraterrestrials" is some kind of Harvard-sponsored affirmation of visitors from space, I should mention that in my two years as a Harvard undergraduate, I claim that I got wired up backwards. Instead of disbelieving everything until it was proven true, Veritas, I believe absolutely everything until it is proven untrue.

However, with that confession in place, my colleague Buckminster Fuller who was kicked out of Harvard twice, used to say, and I believe him, that the world would be saved by professional generalists – thinkers who make it their business to be interested in everything they encounter. In my own thinking I have always been the most respectful and appreciative of English-educated professional generalists such as Gregory Bateson, Kenneth Boulding, and Sir George Trevelyan.

Now comes Stephen Mather-Lees with a double set of virtues as a professional generalist. He is Cambridge-educated and he has trained himself in all aspects of the domain of engineering. So, when he makes the case for something, I am all eyes and ears for his worldly wisdom and profound practicality.

Who could possibly argue with Stephen's call for gaining as much help as possible from beings who are anything from tens of thousands to millions of years ahead of humanity in their science, technology and hard-earned wisdom, while they are seeking to understand and reflect to us our emotional nature and to save us and our beautiful planet from ruination by self destruction (which is well under way).

It is by now self-evident that the spiraling increase of chaos and collapse in the world environment, the world economy, and the world political culture is calling us rapidly towards the need to create conscious human evolution and a new world paradigm to enhance our critically necessary social invention of the human future. The acceleration of human evolution alone makes our new invention strategy a necessity.

We are now learning that humans have been hunting and gathering for millions of years successfully and in some remote areas on Earth, still are. Agriculture and the domestication of animals goes back some ten thousand years. On that success with its attendant ecological and military failures, civilizations have risen and fallen worldwide for some six thousand years. Then after eight hundred years of dark ages in western culture, a renaissance of inventiveness flowered forth, culminating in the development of science for some four hundred years. In our understanding of the universe from the discoveries of Copernicus, born in 1473, to Galileo, to Kepler, to Sir Isaac Newton, to Einstein who died in 1955, our cosmos has been reframed over and over by the progress of science and its derivative world culture. The age of industry and mass culture has circled the world in only two hundred years. And now, since the end of World War II just decades ago, the atomic, electronic, and satellite age was born where now on television and the Internet the whole world's news is everyone's news in real time.

From millions of years to mere decades, human evolution has been accelerating. Now we are at a threshold moment when we must consciously create the human species' future for the common good of all life on Earth for all time, or we will decline through arrogance and ignorance toward our own extinction. At such a moment there are precious few voices comprehensively wise and courageous enough to help reframe our vision of what our best chances are and why we should be seeking the best friendship and guidance from our more mature galactic neighbors. Stephen Mather-Lees is such a voice in his masterwork "Extraterrestrials".

This book has a superb overview of the world's set of malaises, a presentation of reincarnation dissolving our fear of death, an expansive set of questions for our extraterrestrial friends as we ask them to help us mature, and an exciting look ahead into the future of human technology. Each of these alone is a great gift. Yet they are only small jewels in the great crown of human initiation into membership of the whole family of life in the Universe. In this book we are given our crown in a heartfelt summons to our future.

I feel that this call to us is of the magnitude of the English reframing wisdom, generosity and spirit shown to us by the likes of Newton, Darwin, and here in America where this book was written, by the vision of our own revolutionary Thomas Paine in his "Common Sense".

Stephen Mather-Lees is a true visionary of human evolution, offering us the common sense of our species' salvation.

August "Gus" Jaccaci

Thetford, Vermont and New Gloucester, Maine

Preface

Thanks are due to a number of writers and teachers for inspiration and significant contributions to the thinking behind this book. They are too numerous to mention them all, but I would like to single out David Wilcock, Carla Rueckert, Dr. Steven Greer, Michael Mannion, and the late Dr. John Mack of Harvard.

I would also like to thank a number of people for their constructive reviews of the book and their support in general. My wife, Susan Yerburgh, and my long time friends Roger Beeching, Gil and Sue Brinckerhoff, Atira, and Robert Baldridge. Thanks as well to my friend at CSETI, Jack Auman.

Special thanks go to Gus Jaccaci, who as well as reviewing it and agreeing to write a great Foreword, has been very supportive from the beginning.

Introduction

In "Who the Bleep Designed Us?"[1] it was put forward that there are abundant reasons to believe that our evolution was guided by at least one race of advanced extra-terrestrials. That book ended by posing the question "OK, So What Now?". That is a very important question, since it is not merely "interesting" that ETs have been involved with us for hundreds of thousands, if not millions, of years. It was also noted in "Who the Bleep" that this is not by any means a new idea. Writers such as Erik von Daniken[2] and the late Zecharia Sitchin[3] proposed this general scenario some time ago.

This book is an effort to do get the attention of the mainstream of thoughtful readers. It is notable that von Daniken, Sitchin and many others are in fact quite widely read, but it seems that their readership is actually only a fairly narrow demographic, and an even smaller number takes the subject seriously enough to want to do something about it. The reasons for this are not entirely clear, but I feel that the subject is important enough to keep trying, and part of that effort is working to understand the difficulty of getting through to a wider audience.

One important part of this difficulty is the fact that taking the subject seriously requires a significant shift in worldview for the vast majority of people. A shift of this kind was correctly described by the late John Mack in his book "Passport to the Cosmos" as a very threatening thing, which brings up great resistance and denial. This resistance to a serious shift in worldview is extremely difficult to over-

---

[1] My previous book, with the subtitle "A God? ET's? Both? Neither?". You do not need to have read it, but you might find it interesting!
[2] "Chariots of the Gods"
[3] The Earth Chronicles Series

come. A study of the history of the treatment handed out to people who challenged the worldview of their time confirms Dr Mack's observation. Think of Galileo, Giordano Bruno, Nicola Tesla[4], John Mack himself, Wilhelm Reich[5], and of course Jesus of Nazareth. Two of these (Jesus, Bruno) were actually executed by the Powers That Be of their time. Reich died in prison after his work was destroyed by the US government in the 1950's, and all the copies of his books that they could find were burned. Yes, burned, actual book burning in the 20th century, and not in some third world country either, but the supposedly civilized, free speech, First Amendment US of A!

For this reason the first chapter is about how the shifting of a worldview may be made less painful and threatening by taking it in small steps.

Among those whose worldview accepts the existence and presence here of ETs, I suspect that there are many who are concerned about how to deal with the ET presence but feel powerless to make a difference, and I plan to offer those people some ideas that should help.

---

[4] The inventor of alternating current as a means of transmitting electrical power. His life work and battles with the Powers That Be over his patents and recognition for his inventions, is covered well by Wikipedia.

[5] Reich, after early work with Sigmund Freud, made a number of advances in psychiatry, and went on to experiment with what he called "Orgone Energy", arguably the same thing as Zero Point Energy. He ran afoul of the FDA by promoting and selling his Orgone Accumulators without FDA approval. Like many visionaries he was a proud and unyielding man, and did not accept the right of the FDA to shut him down or the authority of the courts. He was eventually convicted of contempt of court. His entry in Wikipedia makes very interesting (and sad) reading.

Anyway, now that we have more compelling reasons to buy into the idea of extra-terrestrial involvement with us, not only in the distant past but here and now, and I think it is time to elaborate on them.

Firstly, it is clear that before long (a few hundred years), we will be perfectly capable of engineering beings such as ourselves starting from the DNA of other mammals, and not long after that, starting with quite primitive animals. At that time, we will not be very far advanced in our own development. After all, we have only been walking upright and making tools for a few hundred thousand years, a mere moment in cosmic time, and have only been sentient in any meaningful way for a lot less than that. That brings me to the question, what does "sentient" mean? Many animal lovers would argue that their pets are sentient, and by a loose definition, arguably they are, as they can be empathetic, protective of their owner, and even self-sacrificing. However for the purposes of this book I would like to propose a more rigorous definition. How about this: A clear sense of self, a wonder and curiosity about our world and the Universe, and a desire to know what it all means. (I do realize that this is not the general dictionary definition, which is woefully vague and in a word, useless. Google "sentient" to see what I mean. Googling "sentient being" is somewhat more useful, but still vague.)

Secondly, the information that we now have implies that many millions, probably hundreds of millions, of sentient races have been around in this galaxy for much longer than we have, and the vast majority of them will therefore have surpassed our technical, as well as our mental, capability a very long time ago[6]. This is because we

---

[6] It is currently thought, based on recent measurements by NASA's Kepler telescope, that there are as many as 50 billion planets in our Milky Way galaxy. Of these, 1 in 200, or 250 million, are in the so-called "Goldilocks Zone" (not too hot, not too cold, but "just right") around their parent star, where conditions support the evolution of

are "the new kids on the block", who have only had technology for a very short time indeed. Given that sentience will still be developing in many of the possible planets, we still have hundreds of thousands of ET races that will have been sentient for times varying from fairly recent (say a hundred or so thousand years) to as much as a billion or more.

So we have to conclude that the ability to engineer the evolutionary process has been present in our galaxy for many millions, even billions, of years. Thus, while it is indeed possible that such engineering did not take place here, it has to be viewed as entirely possible, and indeed quite likely, as you will agree if you read "Who The Bleep Designed Us?".

Thirdly, additional credence has to be given to the idea that our DNA was engineered, because a 2001 report originating from the Human Genome Project (HGP) indicated that the difference between our DNA and that of our closest earthly genetic relatives (that is, the great apes – chimpanzees, orangutans, and gorillas) is of a character that is not the same as the gradual progressive changes that led from the simplest organisms up to the highest mammals. This difference was described in the article as a "sideways insertion" of no fewer than 223 genes! The article was in the magazine Nature in 2001, and hypothesized that the insertion was as a result of a bacterial infection. Shaky at best, but if one is discounting the idea of extraterrestrial interference, one has to come up with something. If bacterial infection is indeed a major factor in evolution, that mechanism would be evident throughout the history of life on earth, and many thousands if not millions of "insertions" would have been found. Zecharia Sitchin wrote an excellent article on this, which is available

---

sentient life. In "Who The Bleep", the numbers used are from previous information which even then yielded a probability of hundreds of thousands of other sentient races in the Milky Way.

on his web site, www.sitchin.com/adam.htm. In this article, researchers from the HGP are quoted as saying that 25 out of the 35 proteins expressed by these genes are unique to humans. It is also possible that the bacterial aspect of this could be that the ETs doing this work use bacteria (or bacteria-like nanobots!) to introduce their DNA changes.

Therefore it behooves us to take a very hard look at ET involvement. "Who the Bleep" proposed that, and also discussed evolution in terms of a persuasive (and verifiable) slant on Intelligent Design, calling it instead "Assisted Evolution". If this approach is taken, the anomalies and gaps in evolutionary theory start to make sense, particularly the very important but neglected one – that we only use a small portion of our mental capacity. This of course is hardly a survival trait! Sitchin's translations of Sumerian tablets recording their history, tell the story that the particular ETs that the Sumerian tablets describe (the Annunaki), bred us for service as extractors of mineral resources for them. This is a rather distasteful idea to us, putting us at a low point on the cosmic totem pole (or if you prefer, the cosmic pecking order); however it does provide a possible explanation for some if not all, of our limitations. Regardless of the validity of Sitchin's work, let us acknowledge that we are in any case at the bottom of the totem pole/pecking order, just because of the very short time (in cosmic terms) that we have been around.

There are numerous books and web sites available that describe alien encounters, abductions, genetic experiments, crop formations, and government's cover-ups of UFO sightings and crashes. Most of these describe events that are so far outside our everyday experience that it is very hard to take them seriously. And indeed many of them may be the results of hallucinations or imaginative publicity-seekers. There is however a large body of work by serious researchers and thoughtful observers, who examine the large numbers of incidents and encounters that cannot be easily dismissed. In his book "Project

Mindshift", Michael Mannion describes a number of instances of encounters that were witnessed by many people, frequently professionals or trained observers of one kind or another. In most cases the military authorities dismissed them as being such things as weather balloons, or atmospheric effects. In one well-documented case reported at the 2007 MUFON Symposium in Denver CO, cars from several police departments in Portage County, Ohio (Southeast of Cleveland) chased a UFO across several counties including part of Pennsylvania for a number of hours. The Air Force officer assigned to the case (apparently with a straight face), assured the officers that they had been chasing the planet Venus! This in spite of the fact that the object was large and had been observed to occlude Venus several times during the chase. One should also note that in order to be chasing Venus, one would need to be travelling West, not East! For those who are shaky on their US geography, you need to go East to get to Pennsylvania from Ohio. And the planets, like the Sun, move from East to West in the sky. Oops.

These dismissals by the "authorities" could be credited if the sighting reports came from ordinary people with little education or training, as indeed many reports do. But they cannot be credited when the sighting reports come from trained observers like police, airline pilots, military pilots, air traffic controllers, and engineers. All of those professionals know perfectly well what weather balloons, atmospheric effects and the planet Venus look like. Michael Mannion's thesis is that many of the world's governments have been well aware of the reality of the ET presence here for over 60 years, in other words at least as long ago as the Roswell incident[7]. His thesis also

---

[7] The most well-known UFO incident, in which a craft of some kind with a non-human crew crashed in the New Mexico desert. It was reported in the local newspaper but later denied and alternative, rather ludicrous explanations were given by the USAF. There are many books on the subject by respected and thorough researchers

maintains that our governments have ever since been muddying the waters in various ways, assisted to an extent by the mainstream media, who relegate such discussion to the Sci Fi channel and similar vehicles.

This "conspiracy" may well exist, and if we are to believe what is revealed in the book "The Day After Roswell" by retired army Colonel Philip Corso, it certainly did and still does. (Col. Corso was retired when he wrote the book; it recounts in great detail how the cover-up was born out of the uncertainty and paranoia of the Cold War. Corso himself was a key player at the highest levels of Army Intelligence and was a part of the cover-up and the feeding of the Roswell craft's technology to various defense contractors from the beginning.) Corso's and Mannion's books are highly recommended reading for anyone interested in how we got here from there, as is the excellent book by Dr. Steven Greer, "Hidden Secrets".

The thesis of this book is, so what? We do not have to go along with the misdirection! We need simply to sidestep it and make contact anyway, holding our own dialogue with the ETs. The evidence that we have indicates that none of the ET races that have been coming here are actually hostile to us in any way (although there is evidence that some are trying to operate in the role of controlling parents). It appears that some are perhaps indifferent observers, others are observers who care but are following a hands-off policy, while at least some of them are operating on agendas that we do not and perhaps cannot - yet - understand. There is however evidence that the US government as well as other governments, has been trying subtly to promote the idea that there are hostile ETs around as part of their "muddy the waters" campaign. It is also sad but true that it is to the advantage of the elite that control the military/industrial/political

---

such as Stanton Friedman, who I have spent some time with and who comes across as being serious and having high integrity.

complex to be able to point to "threats to national security", so that the enormous sums that go into defense continue to do so. Some of the ET's activities are certainly strange – but on reflection, it would be amazing if what they were doing made immediate sense. We should certainly expect them to have a very different outlook on life from us. But rather than being in denial about their presence or being afraid of what they are up to, let us just make it known that we know they are here, and that we want to talk! Even if there is risk involved, surely it is better to find out as much as we can as soon as we can, rather than bury our heads in the sand.

Chapter 1

Levels of Belief

Although many readers will be quite familiar with the concepts being put forward here, many others will be struggling with the extent of the new information and how far it requires them to move from their current worldview.

I would urge those readers who have accepted the ET presence not to skip this chapter, though, because it has a bearing on the difficulty of discussing the subject with family and friends. This is a difficulty that I have encountered a lot, and it is very frustrating! One would think that family and friends would give one a fair hearing, but that worldview thing keeps getting in the way. Another good quote from the New Testament – "A prophet is not without honor save in his own country".

The thesis of this chapter is that for a significant shift in worldview, it is easier for people to move step by step through levels of belief. We should not expect them to take a big leap to get to a big change. The reason for this is the previously discussed phenomenon of the worldview and how difficult it is to get anyone to budge from their version of it. It will be much easier to do it in small steps.

The approach will be to start from a baseline of generally accepted beliefs, then to start thinking about the implications of that baseline. Then it should be possible to start considering the next level.

… almost everyone has direct experience of some form of paranormal event…

The starting point is that almost everyone has direct experience of some form of paranormal event, whether it is "simple" telepathy or whether we use a wider definition which includes any way of accessing information by means other than one of the five "regular" senses. In that case we can include:

- feelings of déjà vu
- the intense feeling of cold that so many people get in the vicinity of a tragedy. A good example of this is Glencoe in Scotland, where a massacre of members of the clan MacDonald took place in 1692.
- the feeling of already knowing someone we have just met
- the hearing of one's name being called when a loved one is in crisis elsewhere
- the connection between identical twins where they "know" what the other is feeling, even over long distances
- the similar "knowing", where a mother knows when a child of hers is in some kind of trouble or pain
- the "knowing" when one is being stared at, that itch between the shoulder blades or the back of the neck

In that the vast majority of us have experienced one or more of the above examples of paranormal activity, it is surprising that so few of us have considered the implications. Which are rather important! They imply very forcefully that telepathy at least is commonplace.

In Scotland and Ireland, it is quite usual for someone to be credited with having "the sight". Native Americans and their cousins in Central and South America have shamans who are credited with healing powers and telepathic abilities, as well as being able to communicate with tribal ancestors. The use of dowsers to locate the best place to dig a well is commonplace. Many people worldwide (myself included) have participated in fire walks (OK, that is not exactly telepathy, but it certainly is paranormal! And mind blowing, especially for a career engineer). Faith healing services are conducted in many places. These beliefs and practices are worldwide in one form or another.

The point is, there is a worldwide level of belief and personal experience in these "basic" forms of unconventional "knowing" and paranormal functioning. This might be termed the first level of belief. Once this is accepted **and pondered**, it becomes easier to consider and accept more advanced forms of paranormal functioning.

It is interesting and informative to note here that the following statement is attributed to Jesus of Nazareth, in John 14.12. "Verily, verily, I say unto you, He that believeth on me, the works that I do shall he do also; and greater works than these shall he do". The various Christian churches have made very little of this, but it is actually very significant. After all, if one is to believe the New Testament (and of course a lot of people do), Jesus healed the sick, changed water into wine, raised Lazarus from the dead, rose himself from the dead, walked on water, and did many other remarkable things. And there he was, stating clearly that we will be able to do all that and more. So do we accept that? Apparently millions of Christians do, as well as accepting the miracles attributed to Mary and the other Saints. Pilgrimages to such places as Lourdes are quite popular. Millions of Muslims attribute miracles to Mohammed. Millions of Buddhists attributes miracles to Buddha. And so on.

… belief in paranormal powers is extremely widespread, much more so than at first glance.

What I am saying is that belief in paranormal powers is extremely widespread, much more so than at first glance. However, there is a curious disconnect between this widespread acceptance that "other people", "gifted people", or "chosen people" have demonstrated these abilities, and the willingness of the general public to look hard at the implications, which are that everyone has at least the potential of one or more kinds of paranormal ability, and that modern science is missing something of vital importance. The mission here, then, is to persuade more people to move beyond that disconnect. We now have a starting point, and if at least some people can move beyond their unconsidered acceptance of their belief in "basic" paranormal functioning and get to a thoughtful examination of the implications of that, it will be possible for them to move on to the next level.

Standing in the way of this is the history of science in the last few hundred years and the insistence of virtually all scientific teaching on the "scientific method", in which any observed result must be reproducible 100% of the time before it can be accepted as "proven". Thus anything that only works some of the time or only for some people, can be and is dismissed by the scientific community.

This is a pity because many experiments with such things as guessing cards, or throws of the dice, or affecting the randomness of random number generators, have yielded significant results. For instance Dr J.B. Rhine at Duke University had a great deal of success with well documented experiments, and his published work spans over 40 years between the mid-1930's and the late 1970's. The success however showed up as statistical differences between the actual results and strictly random results. So even though the cumulative probability of all his results being due to chance was about 10^{-24}, a disappearingly small number, his work has never been accepted by mainstream science, because of course, the card or die was not guessed right every time! A brief word on the statistical part of this: if a single test, say guessing 100 tosses of coins, show 52 correct and 48 wrong, that on its own is not significant. But if that test is repeated 100 times and each shows the same deviation from chance, the probability of that being chance is now 48/52 to the power of 100, or about 0.0003. Since there were so many of Rhine's experiments and he so consistently got good results, the combined probability came to about 0.000000000000000000000001, which is close to 48/52 to the 700th power, just to give you an idea.

Another important example of the "basic" paranormal effects was researched extensively by Cleve Backster, who found that plants and even biological cultures such as yogurt, reacted with nervous shock responses as detected by an instrument similar to a lie detector, not only when actually harmed by such methods as cutting or burning, but also when the experimenter only thought about harming them by whatever means! Again, this is not 100% reproducible. But it is certainly a real and remarkable effect!

The same problem with reproducibility has dogged the pursuit of Cold Fusion and Zero Point Energy. Quite a number of people in universities and laboratories all over the world have seen amazing results in both areas but are typically unable to reproduce them reliably on demand. It appears that the mental state of the experimenter plays an important part in the result, which of course is not popular with followers of the scientific method. But that in itself, that the mental state of the experimenter plays a part, is both amazing and highly significant.

Mainstream science has had to go along with the concept of the role of the observer at the sub-atomic level because it is impossible to find out what is going on there without interfering in some way with the particles involved. Actually, the functioning of the transistors in our televisions, cell phones, computers and communication equipment relies on statistics! Those statistics that explain the functioning of semiconductor devices were derived from work by Enrico Fermi and Paul Dirac, and are called Fermi-Dirac statistics. Both Google and Wikipedia will yield good information on "Fermi-Dirac" and "Heisenberg's Uncertainty Principle", or if you have the stomach (and doctoral level math) for it, get into Quantum Mechanics!

However at least so far, that acceptance has not been extended to the macro level of such things as Dr Rhine's work, or indeed any of the "basic" paranormal effects. A more recent example of telepathy was the experiment performed by Edgar Mitchell on the Apollo 14 mission, where he not only had an extremely high accuracy guessing cards, but demonstrated that he was getting the information effectively instantaneously, and certainly many times faster than light speed[8].

Once one has looked at the implications of the "basic" level, the next level of belief to be considered might involve such things as remote viewing and contact with loved ones who have passed on. There are a number of talented psychic mediums who regularly draw quite large crowds and are able to convince members of their audiences that they have messages from "the other side". As mentioned earlier, the CIA certainly believed enough in remote viewing to invest many years and a lot of money into it, and several of the people who worked in that program are now offering courses in it. So a significant number of people have already reached that level of belief. Again, even the most gifted psychics only claim about 80% accuracy. It might be observed that there is no profession whose members claim to be right in their own fields 100% of the time, since that is clearly ridiculous. But because a psychic or remote viewer is right "only" 80% of the time, their work is dismissed by mainstream science!

---

[8] The way this was done was to carefully time the moment that the information was "transmitted" and when Edgar Mitchell's response was received. Light takes about 1.3 seconds to get to the moon (just under 240,000 miles at 186,000 miles per second). So if the effect was light speed limited, it would take at least 2.6 seconds (plus Edgar's reaction time) for a response to be received. In fact the total time was much closer to 1.5 seconds, indicating that Edgar was getting the information effectively instantaneously, or anyway much, much faster than light speed.

Once the efficacy of remote viewing is accepted, the next level would involve examining what kinds of things remote viewers have seen. It turns out that they have seen (and had confirmed) seeing things not only remote from themselves and not known to the person feeding them the reference information (usually just a number which refers to a place, event or person known only by a computer), but also remote in time. They have seen ETs in spacecraft. They have observed ET facilities on the Moon and on Mars. So if we only attribute the same 80% accuracy to remote viewers as we do to psychics, they have still seen a lot a very remarkable things!

It is also reasonable to consider that psychic functioning can extend to such things as channeling, where a person connects to a being (not necessarily human) who exists in another dimension or density (one example of this is of course getting messages from people who have died). Examples of channeling from extra-terrestrial beings in other dimensions of existence are found in the Ra material channeled by Carla Rueckert, and in the Seth material channeled by Jane Roberts. Training in this skill is also available on the Internet.

Once there is acceptance of remote viewing and communication with beings that are outside our reality, some of the more mysterious reports of UFO behavior start to make sense. I am referring to the reported ability of ET craft to appear and disappear instantly, to perform maneuvers like right angle turns at very high speeds, to change shape and size, and to be visible to some people but not others. Steven Greer reports that on many occasions he and others have seen ET craft that were partly materialized, in other words they were transparent, but the craft and its crew were nonetheless clearly visible. In some cases UFOs are observed on military and commercial radar, in other cases they are not.

In the past it has been reports of effects like these that have enabled skeptics to dismiss those reports, because, of course, such effects are "impossible".

It is important at this time to open our minds to a reality in which our present science is inadequate to account for what is going on.

It is important at this time to open our minds to a reality in which our present science is inadequate to account for what is going on. This has to a greater or lesser extent always been necessary, as new information comes forward which challenges the prevailing worldview. It is sad but true that the scientific establishment and/or the religious one, has always and without exception fiercely fought each new important idea. Examples of this are:

- Galileo's proposal that the Earth orbited the Sun instead of the other way round. The Church forced him to recant.
- Giordano Bruno was actually executed by the Church for the same thing.
- Darwin's theory of Evolution was fought bitterly and some Christian sects are fighting it even today.
- The Wright brothers' first flights, even though they were witnessed by hundreds of people, were denied by none other than Scientific American and the New York Herald, for literally years after they happened.

This opening of the mind, at least to the possibility of paranormal effects, is a vital step towards making contact with our ET neighbors and starting a dialogue. The reason why should be obvious. We are dealing with multiple races of beings whose technologies are vastly ahead of ours. Of course they will be able to do things that we do not understand and cannot explain with our current understanding of the way the world works! Remember, as discussed in the Introduction, their technologies are, probably in almost every case, hundreds of thousands of years ahead of ours. If we want to have everything that we experience to fit comfortably into the prevailing worldview, we will be doomed to stick right where we are. And sadly, that will be exactly what most people will want to do. A worldview for most people is a comfort zone, and they will resist any attempt to move them out of it. But how much more interesting and exciting it is to have a worldview where there are a semi-infinite number of things to be discovered and experienced, and to be not only willing but expecting to be regularly surprised!

This undertaking calls for a childlike form of wonder and curiosity. And why would that be a bad thing? The study of science and physics actually requires it – unfortunately most scientists and physicists have lost that wonder and willingness to be open to new ideas and information. A sense of wonder and willingness to entertain new ideas does not of course mean that we have to throw away our critical faculties, by the way. It simply means that we need to be open-minded.

The demands by the universities and institutions to conform, to publish "safe" material in order to get tenure, and to stay away from controversial topics, are very real and powerful. However all of that, as one of my old schoolteachers used to say, "is an explanation, not an excuse".

Chapter 2

Why should we be talking to the ETs?

"Minds are like parachutes – they only function when they are open." James Dewar

Abstract: Almost all of us were brought up to believe that talking to God is praying. However what we were taught is more like talking at God rather than having a dialogue. Whether or not there was an original Creator, the entities responsible for most of the actions that we have attributed to God, are members of one or more races of ETs! Therefore, we need to re-think what praying should be in these times. In meditation, some of us learn to listen as well as talk, which is a big improvement. But, at this time and in this arena, we need to ask questions and get answers! It is time that "our" ETs' role shifted from an assisting and perhaps controlling one, to a mentoring one. Whether we are a part of some long term plan or a series of arbitrary experiments by a succession of ET races, it is time to start participating. In other words, it is time to get out of cosmic kindergarten and into first grade. This metaphor is particularly apt. In kindergarten we are almost entirely fed with information and instructed what to do all the time. In first grade, we are encouraged to start interacting with our teachers, and to ask questions.

There are a number of organizations who study UFO sightings and work to get information on them from the government using the Freedom of Information Act. This has served a useful purpose in that it has become clear that the government knows a great deal more than it has told us about UFO encounters and UFO technology. After stalling for years and many court battles, a government agency blandly released hundreds of pages of data, with virtually the entire text blacked out "for national security reasons". But of course, Nothing Is Going On. One has to laugh (or cry?) at this manifestation of our tax dollars at work.

… although all those things are interesting, they are not yielding any useful information about the ETs themselves, or about their agenda, or about their technology.

The thing is, the point has now been made in spades. It is no longer useful to hack away at the secrecy around the Roswell incident, or the Betty and Barney Hill abduction, or the great Ohio-Pennsylvania UFO chase, or the Chicago O'Hare sighting over the United Airlines hangar. It is not even useful to obsess over crop formations and try to interpret them, although they certainly are fascinating. The messages such as they are, are either mathematical in nature (one recent one very cleverly displayed the value of pi to ten decimal places, for instance, and another spelled out the famous Euler equation "e^(i*pi)=0" in ASCII code), or to do with indicating future dates by means of the relative positions of the planets. The fact is that although all those things are interesting, they are not yielding any useful information about the ETs themselves, or about their agenda, or about their technology. At best, the crop formations are letting us know that our intelligence is being calibrated by examining our reaction to, and interpretation of, their information content.

Plenty of scholarly work has been done with regard to ET involvement in our development, notably by the late Zecharia Sitchin and Paul von Ward. Their work is based on the common thread of interaction and interbreeding with "gods from the sky" that shows up in almost all of the most ancient writings worldwide. (Sumerian, Mayan, Egyptian, Greek, Roman, Hindu, Chinese, Nordic, Celtic, Native North, Central and South American, various African, etc.)

While all of that makes fascinating reading, it is time to look forward, concentrating on what is happening today and what we should be doing in a pro-active sense. The holders of the genuine information which can make things clear to us, are of course the ETs themselves. Since it is regrettably obvious that we cannot rely on our governments, we have to take matters into our own hands and contact the ETs under our own initiative.

How do we do this? That is a very good question that is addressed in Chapter 4. For now, let us concentrate on why it is important to do it at all.

The state of this world is a very perilous one. The violence stemming from fear and intolerance which has characterized humanity ever since we were hunter-gatherers, is now infinitely more destructive than it used to be, because the weapons we are using and the means of deploying them, are so much more effective. Now that the nuclear genie is out of its bottle, and the bio-weapons genies are emerging from their bottles, it is more important than ever that the underlying causes of violence are addressed.

Can we solve all of our problems on our own? Well, maybe. And, maybe not! Do we dare take the risk that we can do it without help? Surely we should be pursuing all possible avenues.

Can we solve all of our problems on our own? Well, maybe. And, maybe not! Do we dare take the risk that we can do it without help? Surely we should be pursuing all possible avenues. It seems absolutely undeniable that the information and guidance we need is out there. Some of the information we get is likely to be incorrect or misleading, or just plain confusing. That does not mean that it is not worth getting, because at least some of what we get will be genuine.

But, and this is very important, the genuine article will come with evidence that is verifiable. We will know it "by the fruit it bears", by the degree of openness of the source, and by the degree to which any technology that it comes with is benign. One thing above all that characterizes a sincere source of information on any subject, is that it is only too glad to provide proof and/or means of verification. It is happy to go into detail. The sources that we should absolutely distrust, are those that say "trust me", "what, do think I'm lying", "how dare you question my good faith", etc. I am convinced that this is a universal truth.

It is also true that much if not all of the information we need, can be found inside ourselves by meditation and general self–improvement. I am by no means proposing that those means should be set aside. They rather complement the external search and give us the means to cross-check.

Unfortunately the majority of the world's population is engaged in simple survival, and a high percentage is also enmeshed in the bigotry and intolerance being taught them by self-serving elites and priesthoods. This teaching of fundamentalist ideas is not limited to extremist Islam, but is widespread in the United States as well where many believe that every word in the Bible is literally true (at least, those words that match their worldview), and that everyone that does not share their beliefs is going straight to Hell. This Christian extremism manifests in the way that several states keep trying to put Creationism on an equal footing with Darwinian Evolution in their schools, and extremist views of many kinds are plentiful on the Internet, on radio and television, and in the political arena.

It is unfortunate that extremists of all sorts because of the energy that comes from their deeply held convictions, can be quite successful at getting into political office (for instance James Watt, the Secretary of the Interior under Ronald Reagan). Some of these extremists including Watt, have gone on record as saying that global warming and destruction of the Earth's forests, rivers, fisheries etc. do not matter because the Second Coming is happening any day now, making it all moot.

What this all implies is that it is very important to increase the number of people who take the plight of our planet seriously, and that the idea of looking to our galactic neighbors for help is a worthwhile one.

One huge leg up that the ETs can give us is technology - to generate cheap and clean power, and to feed our population without destroying our planet.

One huge leg up that the ETs can give us is technology - to generate cheap and clean power, and to feed our population without destroying our planet. The point is that the ET civilizations that are out there have clearly come successfully through their version of the crisis that we are currently in. Somehow they managed to feed their population, to limit growth, to come up with power sources that did not ruin their environments, and also to travel around the galaxy.

This means that they figured out how to grow enough food without using enormous amounts of energy, and without the mountains of fertilizers and pesticides that we are presently using to maximize yield per acre. Our "modern" techniques of producing food are:

- polluting our water supply
- depleting the topsoil where crops are intensively farmed
- exhausting our water supply (rivers all over the world are no longer reaching the sea for part of the year)
- creating and expanding deserts and dust bowls
- creating "dead zones" in the oceans at the mouths of many rivers from fertilizer runoff

- steadily clearing huge areas of forest because of the need for more arable land
- wiping out fishing grounds from over-fishing

Our appetite for energy, which at present we get almost exclusively from fossil fuels – oil, coal and natural gas, is causing us to:

- destroy important habitat and other natural resources by drilling for oil in such places as the Amazon rainforest, the Niger delta, and the North Slope of Alaska. Not to mention the Gulf of Mexico! Many of the countries of the Middle East have already suffered widespread devastation while their rulers (not their people) have become wealthy.
- ruin large areas of places like West Virginia with techniques such as "mountain top removal" to get coal from near the surface. The tar sands of Canada are another case in point.
- pollute aquifers with natural gas (methane) by breaking up gas-bearing strata to release the gas, which then finds its way into the aquifers and hence to residential drinking water. This results in the awful phenomenon of people being able to light up their household faucets.

It is what Al Gore would call "an unfortunate truth" that those forms of alternative energy which are being slowly and painfully introduced now, are far from environmentally benign, although they are certainly an improvement over fossil fuels. Wind, Solar, Electric cars, all use a lot of energy in the manufacturing process and use expensive and rare materials, as well as creating significant pollution from manufacturing waste. So while they may buy us a little time, they are a long way from being the kind of solution that we really need.

A technological leg up from more advanced civilizations will enable us to better educate ourselves and make use of the rest of the new knowledge that they can bring us, to broaden our understanding of each other and our Universe. Our own history shows that once a population is adequately fed and educated (and women are given equal rights), the birth rate drops to a sustainable level. The latter is in many ways the most important result that we need, in that the world is already seriously over-populated, and it will become more so even if we were able to implement a sustainable birth rate today. (This is because there will be a significant increase in population from the future children of the young people already born.) It is clear that as well as knowing how to grow more food, we need to come up with and implement a set of programs that will bring the population down to a sustainable level. Once people are educated enough and incentives are in place to limit children to 2 or less per family, the population will start to come down. Coupled with education we also need a new economic system which does not depend on constant growth and ever increasing consumption. It is ironic that growth and consumption are held to be positives in mainstream economics, while in the medical field growths are at least ugly and often malignant, and consumption is another name for tuberculosis!

Can we achieve all the above goals by ourselves? Perhaps, eventually. But we are already seeing that there is tremendous inertia in the system, and it seems inevitable that some disastrous event or events will force us to do it the hard way, via a catastrophic reduction in the population from disease and/or famine and/or war over our increasingly scarce resources. Realistically, it will take many generations to get enough of our billions sufficiently well fed and educated to turn the population engine around. Even in the so-called civilized countries it will take a long time to change our economic paradigm from one of consumption and growth to one of balance and sustainability. That is why we really need to be looking for some help. Even then it will not be easy, since so much power is vested in people who benefit financially from the system as it is currently working. But the longer we wait, the harder it is going to be, and the greater the likelihood of various unpleasant events, of which the rising sea level from global warming is probably the least painful.

The fact is, we do not have the right to take the chance that we can somehow muddle through by ourselves. The safety and wellbeing of our world and of our grandchildren are a whole lot more important than our pride!

Help is clearly available "out there". So let's ask for it! If the answer is No, we will be no worse off than we are now.

Chapter 3

And who are "They", anyway?

There are not surprisingly various estimates of the number of ET races visiting our planet at this time. To the extent that there is a consensus, the number appears to be around 20, although I have heard estimates as high as the thousands. In light of the recent observations by NASA's Kepler telescope which indicate that there are of the order of 250 million planets in the Milky Way capable of supporting intelligent life, 20 might well be a low estimate.

Among the places of origin which have been identified, with their approximate distances from us in light years, are:

- The Pleiades   440
  (a cluster of 254 stars, with many more than that number of planets)
- Zeta Reticuli   39
- Orion   88 – 576
  (more than one race from different star systems)
- Sirius A   8.6
- Sirius B   8.6
- Arcturus   37
- Betelgeuse   520
- Aldebaran   65
- Lyra   25
- Altair   16.7
- Alpha Centauri   4.4
- Cassiopeia   19
- Alcyone   430
- Taygeta   440

All of the above are well-known star systems relatively close to us (within a few hundred light years). To those can be added:

- Nibiru – according to Zecharia Sitchin's interpretation of Sumerian writings, this is a planet, a member of our Solar System with an extremely eccentric orbit with a period of about 3600 years. It presumably has its own source of heat, either from some unknown technology or an accompanying dwarf star. If the latter, it would have to be a very low intensity one, emitting in the infra-red part of the spectrum, to account for us not seeing it already. Or perhaps it has been observed and that information is being kept from us.
- Other galaxies. If FTL travel is indeed possible, then so is traffic right across our galaxy, as well as between galaxies. There are those that think we are being visited from Andromeda as well as other galaxies in the supercluster to which we and Andromeda belong. (The Milky Way lies just along the outer fringe of the Virgo supercluster of galaxies. The Virgo supercluster contains 150 large galaxies and over a thousand smaller galaxies.)
- Other "dimensions". This gets us into the realm of philosophy and spirituality. Which appears to be necessary! Time will tell.

Various writers (see the references at the end of the book – Whitley Strieber and Jim Sparks are good examples) described the appearance of these races. They fall into several categories, but the most commonly reported are these:

- The "Greys". The best known and most often reported. Diminutive compared to us, proportionally larger head, with large almond shaped dark eyes which may actually be a form of integrated goggles to protect their actual eyes from our sunlight and/or provide night vision. These have been

described as coming in two sizes, around 3 feet in height and about 5 feet in height. Some writers have stated that the smaller ones are merely low level operators and may have been bred for that purpose.
- Humanoids. Able to mingle with us undetected. This may be because their natural appearance is quite different and they can change it, or because they are in a real sense our relatives in that we have ancestors in common with them. If the latter, and we buy the idea that our evolution was "assisted", it may well imply that humanity is already widespread, at least in this galaxy. There are many who think so.

So far we have where they probably come from, and what they are reported to look like. That gives us some useful information, but does not tell us much about who they are. In other words we may know that a person is from Detroit, and have a full physical description including fingerprints. However this person may be a Saint, a stockbroker, or an axe murderer. They may be straight or gay, religious or atheist, Democrat or Republican, and so on. We have also to bear in mind that there are wide spectra of personality, inclination, and ability in any one race, certainly in ours, and that there could be and probably are, factions within any given race of ETs that have different agendas as far as humanity is concerned. However, at this time and in this context, "Who They Are" is something that we have to think in terms of for each race of ETs as a whole. Then when we get to know them, we can start to distinguish different groups within each race with their own characteristics.

I believe that it is important to point out that strange-appearing ETs, if they are visiting us, are no more or less likely to be benign than ETs that are entirely humanoid, and so we must be prepared to interact in a friendly fashion with any ETs that we may encounter whatever they look like. This will not be easy considering our history of fear and discrimination in dealing with members of our own race who look, behave, or believe even slightly differently from ourselves. When the difference in appearance is more extreme than anything that we could experience on Earth, and the behavior and customs are based on a culture that is wildly different from ours, it will indeed be a challenge. But it is one that must be faced and overcome.

If we imagine our own race exploring this galaxy, there may be different groups of us doing so (for example, at this time, that might be the United States, Russia, or China). While it is possible that more than one group may arrive at the same planet at the same time and start interacting with pre-sentient beings there in different ways and operating with different agendas, that is an unlikely (but certainly not impossible) scenario. People who have read any of Zecharia Sitchin's books will remember that the Sumerian records show that there was conflict between rival groups of the Annunaki. One group viewed humans just as workers who had to be told what to do, while the other group favored assisting our development as a species. In our various mythologies, which Sitchin and others point out are eerily similar in most of the ancient writings all over the world, there is usually rivalry between the various gods, whether the myths originated in Sumer, Egypt, Greece, Rome, Scandinavia, the Americas, or India.

So how do we tackle this? The guiding principle has to be that we evaluate them in the same way that we evaluate each other (when we are using our common sense instead of our emotions and prejudices, that is). This is that we base our evaluation not only on what they say, but more importantly, on what they do. For that reason we need to be very cautious about information that comes via telepathy or channeling, via crop circles, via books, via the Internet, via the news media, or via our governments. The trouble with this of course is that virtually all of our information comes to us via one of those means! Whether information comes to us directly via contact, personal observation or intuitive messages, it needs to be evaluated carefully. It should be checked and shared with others so that it can be discussed. Confirmation and feedback are vital.

The high degree of difficulty of this exercise is matched by its extreme importance. In other words, just because it is difficult, it is a really bad idea to throw up our hands and let matters take their course. Again, we have a responsibility to our grandchildren!

A complicating factor in discovering who the visiting ETs are, is that many ET encounters and UFO incidents do not in fact involve ETs at all, but are either deliberate attempts by the military to muddy the waters, or field testing of secret craft and technologies that the military are developing. This makes it even more difficult for us to discover and understand who our visitors really are, and makes it all the more important to evaluate any being that we encounter and interact with, on their actions rather than their words, and to be very aware that they may not be what they appear to be. My study of the abduction literature has definitely given me the feeling in many cases of earthly military operations (the Betty and Barney Hill abduction[9] is a prime example of this), and the animal mutilations definitely do. Apart from anything else, it is very hard to believe that any intelligent beings would travel across hundreds of light years to cut up cats and cows! Or for that matter, to snatch people from their beds or cars and give them medical exams, when all the information they could possibly need would be readily available to them with the kind of technology they must have.

---

[9] Betty and Barney Hill were apparently abducted in the White Mountain area of New Hampshire in September 1961. Their story has been told in a book "The Interrupted Journey", and a 1975 movie "The UFO Incident". The Wikipedia entry on them is very detailed and informative.

The story being put around by some that the ETs are here to use our DNA to freshen up their own race's DNA and/or are breeding human/ET hybrids, is not only preposterous, but dumb. ETs that were capable of engineering our DNA to make us what we are, would hardly need to get some of it back! And even if the ones doing it are not the ones that engineered us, their technologies would be so far advanced in order to do what they do and travel where and how they travel, that (a) they would certainly be quite capable of ensuring that their own DNA did not deteriorate, and (b) even if it did, they would have no problem making their own corrections without the need to leave their own planet.

It is worth repeating here that if the ETs visiting us had hostile intentions, we would have known it very early on, at least in the 1940's and probably earlier, and the game would have been over very shortly afterwards. It is actually encouraging to note that after the Roswell craft was brought down, either by our hostile fire or more probably, because our high powered radar interfered unexpectedly with their navigation systems, they took no action to avenge the destruction of their craft. If the roles were reversed, do we think that our military would not quickly and violently respond?

Chapter 4

How should we make contact?

"Reach out and touch someone" Commercial sound bite

Abstract: This chapter will talk about the various kinds of contact that have been reported, distinguishing between contacts that are clearly extraterrestrial in nature, and contacts that are more likely earthly in origin.

Well, the ETs that are visiting us do not have an Email address or a listed phone number (OK, OK, they might have, but I really do not think so)! It is actually quite a difficult question. The approach of just looking up into the sky and waving a UFO down if you see one, is what one might call a low-probability method, although it could work eventually. When we congregate at conferences devoted to this and related subjects, let us devote a session to doing just that. There have been efforts at this, notably by Ed Grimsley and others, who have found that flashing a lamp of some kind will get a response from observed craft, usually a light that blinks in the same sequence. Needless to say, we have to do a whole lot better than that! While it is evidence of a kind, it is not exactly a conversation!

One difficulty that we have is that while we know that all of the visiting ETs must have vastly more powerful technology than we have, we do not know just how much more powerful it is. There is a lot of territory between assuming that only technologies known to us are useful, as in the rather naïve SETI project (not to be confused with CSETI[10]), which is only looking for radio waves, and assuming

---

[10] For those who are relatively new to this stuff, SETI is a government-funded Search for Extraterrestrial Intelligence using radio telescopes to search the skies for radio signals that indicate an intelligent

that our ET visitors are all-powerful and all-knowing, and "obviously" know what we are thinking and doing, so that we don't really need to do anything to make contact. We do know that there are many, many instances of some kind of UFO contact which were not solicited by the human contactees, at least not consciously, which makes me wonder how those people were chosen. If we can figure that out, we can then perhaps influence the choosing process and change it from a random one, to one where the ones chosen are prepared and ready to interact. Many of the involuntary ones report that they were abducted and/or had scary experiences regularly for years, sometimes from childhood on, before they came to terms with ET contact (notably Whitley Strieber, Lisette Larkins and Jim Sparks, who have all published very articulate books on their experiences). The late John Mack, Michael Mannion, and Budd Hopkins have all published work that carried detailed accounts from a number of experiencers. It is disturbing to note that all too many of the so-called "abduction" experiences have the flavor (or should I say odor) of a military operation designed to frighten people and promote the idea of hostile and unpleasant ETs.

Two people are worth mentioning here who see ET contact as a positive experience. The first is Robert Dean, who has talked about it in a number of interviews and presentations in recent years. His viewpoint comes from his military service in NATO at SHAPE headquarters in Europe. He feels that it is not only ethical but vital that the people be told the truth about contact with various ET races since World War II, so that we can start to take our place in Galactic society.

The second is Dr. Steven Greer the founder of CSETI, who has had positive and spiritual contact with extraterrestrials since childhood.

---

source, while CSETI is a private non-profit, Center for the Study of Extraterrestrial Intelligence, which knows that the ETs are already here and works to contact them directly, as well as working to persuade the US government to disclose what it knows about ETs.

His book "Hidden Secrets" has a lot of detail on this, and in recent years he has had a lot of success in getting together groups of people and using various methods of signaling, inviting ETs to make themselves known and seen[11]. The group approach has of course a more powerful mental signal than a single person, some say much more so than a simple multiple by the group size. I have myself participated in one of these trainings and contact was undoubtedly made. (Check out www.cseti.org for more on this.)

It seems then that there are a couple of possibilities for how experiencers are "chosen".

- The actual experiencers must be extraordinary in some way that we do not yet understand.

  Jim Sparks in his book "The Keepers" reports that he was shown a number of images of members of his blood line going back to the Romans and even beyond to the Stone Age, from which we could draw the inference that the ETs have been working on or at least keeping track of his family for a long time. This could imply that only those family lines that were chosen way back to have their DNA worked on, will have one or more members that the ETs can communicate with. However, while it does seem clear that ETs have been working on or at least keeping track of human DNA for many millennia at least, it does not make sense to conclude that they have been breeding for just a small number of individuals to work with today. With their advanced bio-engineering capability, it is also hard to believe that they would need dozens or even hundreds of generations to accomplish what for them is surely a simple task. So do we disbelieve Jim Sparks? I for one do not think so – I have met and talked with Jim Sparks and have read his book "The Keepers". I believe that there are two possibilities – first, that

---

[11] The CSETI Ambassador trainings, see www.cseti.org.

it makes more sense to position his experience as being outside the mainstream of ET behavior, an individual research project by a small group perhaps, or experimentation by a group that is still relatively new to advanced bio-engineering, that is only a few hundred years more advanced than we are! Second, that those ETs were merely showing Jim Sparks that they had been involved with human evolution and that Jim's family history had been recorded in detail for a long way back. In Jim's book, "The Keepers", it was not stated explicitly by the ETs that he had been "chosen".

Again, time will tell.

- The choosing process is random.

This does seem to fit the facts to this extent: if you are choosing a few hundred or even many thousands from a population of well over 6 billion, you are very unlikely to pick a top scientist or a UFO researcher! But it is nevertheless hard to believe since the visiting ETs must have access to our news media, to published works in print and on the Internet, and to the work of the many researchers. They would also undoubtedly be aware of the many yearly conferences on the various aspects of the ET presence here. It is undeniable that almost anyone working in the UFO research field would love to have personal experience of contact.

- The ETs are avoiding scientists and committed researchers for reasons that are unclear.

This also fits the facts as we know them, with the interesting exceptions of Robert Dean and Steven Greer, mentioned above. We can speculate, however! One answer might be that the ETs have reached a (secret) agreement with the Earthly powers that be, to limit their activities to "ordinary" people. Another might be that there is a galactic quarantine on this planet just because we are so screwed up, and that only limited and devious contact is allowed. This is the thesis of Alfred Webre in his book "Exopolitics".

- They are waiting until we are smart enough and mature enough to initiate from our end.

    Since the ETs obviously know what communications technology we have, it is fair to assume that if they wanted to initiate the kind of dialogue that I am talking about here, they would be transmitting loud and clear. The situation on our planet is nowhere near one where we can claim *as a race* that we are mature enough. But perhaps some of us are mature enough as individuals, and those few need to get smart enough to get the dialogue going. Indeed, it appears that quite a few of us already (and quietly) are doing just that. Steven Greer claims that there are thousands of people worldwide pursuing contact with various ETs.

If we go with the fourth option, which in my view is both the most attractive and the most plausible, the way that we can get smarter is to listen to those experiencers who are willing to talk, those who are not intimidated by the ridicule and the put-downs that come from the majority of the scientific community (who themselves in almost every case, have no experiences, only opinions). Probably the best example of such an experiencer is Steven Greer, and I do recommend his book "Hidden Secrets". From his work and also my own experience, it does seem that it is important to develop some skill in meditation and remote viewing. For most people, it will take a fairly intensive course to develop these techniques to a useful degree. Such courses are offered by, among others, the FarSight Institute (www.farsight.org), Western Institute of Remote Viewing (www.remoteviewers.com), PSI Tech (www.remoteviewing.com), and Straightline Remote Sensing (www.viewzone.com).

We can consider the plausibility of the first option (development of telepathic powers of communication over a number of generations), as being in some cases combined with the fourth option. To quote from Lisette Larkins' book "Calling on Extraterrestrials":

> "... the claims of tens of thousands or even millions of otherwise normal people whose own lives have been changed, touched, or impacted by contact phenomena are not usually taken into consideration as part of the evidence. And so, the common person, the one in the trenches, the softest voice in the crowd who usually can offer the most insight through the evidence of their own experience, is the least likely to be consulted about what the phenomenon means to the experiencer or to society as a whole.
>
> Experiencers represent millions of silenced witnesses, shut down by the culture, the experts, and the scientists who don't want to hear from them. Instead, scientists, researchers and others with secondhand knowledge and no personal experience are queried. The media want to know what the evidence shows and the experts are well versed in measuring the evidence, but only as a result of concepts with which they are trained or familiar."

In another part of the book, however, she makes the claim that the ETs (the ones that she is in contact with, anyway) are ready and willing to work with us at any time – all we need is the intention and attention, plus a readiness to notice their perhaps bizarre and unexpected methods of making themselves known to us. This notion of the ETs metaphorically sitting by their telephones and waiting for us to call, and then devising weird and indirect methods to contact us, makes little sense, and is belied by my own experience, and also in my opinion by the experience of Lisette Larkins herself. By her own account, she was not in the least focused on soliciting ET contact at the time of her first experiences, but was rather totally embroiled in a major crisis of her personal life.

There is perhaps also a logistical problem – how many ETs are actually waiting for our call? The pattern of UFO sightings and other incidents, suggests that a number of ET races, perhaps of the order of 20 as discussed in Chapter 3, are involved here at this time. One wonders how many beings are actually available for contact, and how many of those are willing to make themselves visible and/or available for a dialogue, telepathic or otherwise. When you consider the total land area of the globe, and that at any moment, most people are indoors working, eating or sleeping rather than being outside noticing anything strange in the sky, it does indeed become a logistical problem. Most of us when outside our homes are playing outdoor sports, basking in the sunshine, reading a book while we sip on the drink of our choice, gardening, or driving somewhere with our attention on the road (hopefully) rather than the small amount of sky visible through the windshield. How many are just waiting for humans to solicit contact, we wonder? It may not be very many, and they may also be quite picky in terms of the attitude and intentions of the human(s) attempting contact.

That scenario leaves aside the previously mentioned idea that we are likely to under a form of quarantine, which would severely limit how many ETs are permitted to be here and would place many restrictions on their activities.

A review of the literature on the subject of ET contact delivers quite a confusing picture. Writers come from a wide variety of backgrounds and attitudes – here is a list.

1. Experiencers. These range over a gamut from CE-1, Close Encounters of the first kind or simple UFO sightings at relatively close range, all the way to CE-5, Close Encounters of the fifth kind, where contact is solicited and unambiguously acknowledged, as in those reported by Steven Greer[12].

---

[12] CE-1 is defined as a sighting of one or more UFOs. CE-2 is defined as a UFO observation plus one or more physical effects such as radiation, markings or depressions on the ground, lost time, etc. CE-3 (dramatized in the movie of that name) is defined as observation of a

2. Researchers who have not themselves had ET contact. Very few of these take an objective approach.
3. Researchers who seriously study the evidence and keep their minds open. There are, sadly, only a few of these.
4. People who have been members of government agencies and claim to have information on cover-ups and other forms of misinformation. Some of these appear to be legitimate, while others appear to be part of the disinformation, muddy the waters campaign.
5. Skeptics and debunkers.
6. "Official" government reports, all of the Nothing Is Going On variety.
7. People who have made up some kind of ET contact story for their own reasons.

Given that many of these are either distorted by the writers' belief system, their desire to muddy the waters, or their desire for notoriety, it is very hard to sort through it all and find the core of truth. This is all the more reason that we should be seeking our own experiences. But given the probably small number of ETs that are here and are making themselves available for contact, our chances do not appear to be good!

UFO plus "animate beings". CE-4 is defined as direct contact with ETs including abduction or voluntary boarding of an ET craft. CE-5 is defined as contact with ETs that has been consciously and deliberately solicited by human beings.

The primary ability that shows up in almost all credible accounts is that of telepathy.

This makes it vital to work with those that have had experiences with ETs and discover what it is that makes them targets for ET contact. The primary ability that shows up in almost all credible accounts is that of telepathy. This ability must either be an especially powerful one, or one that is different in character in some way from the forms of telepathy that so many of us have experienced – that of twins being aware of each other's thoughts, mothers knowing when their children are hurt or in trouble, the sense of being stared at, knowing who is calling when the phone rings, etc. As one reads the many publications by and about ET telepathy, one is struck by the way it is described. It sounds quite different from the vague and transitory (although quite unmistakable) feelings that many of us can attest to. We read of compelling pressure and very specific words, even detailed instructions. We also read of people having a sudden and intense desire to go to a specific place, where contact with the ETs then occurs.

Both the special case of telepathy that is called remote viewing, and the ability to guess cards or dice, do not seem to correlate at all with ET contact, except in the case of Steven Greer and the people who have done his trainings, and some of those people who were involved with the government remote viewing research, notably Courtney Brown. Mr. Brown was not originally involved with the government research, but was trained by someone who was, and has had considerable ET contact using those techniques (see his books "Cosmic Voyage" and "The Science of Remote Viewing"). There seem to be few other cases where someone who was adept at either of those things, has reported contact, telepathic or otherwise. If telepathy is the method of communication, and remembering that it has been established (see Chapter 1) that information transfer is not subject to any speed of light limitation, there is no requirement that the

ETs communicating with us are anywhere close to us! They could be in the same room as we are or sitting in their own home on their own planet elsewhere in the Milky Way or even another galaxy.

If we take a wide definition of telepathy which includes any way of accessing information by means other than one of the five "regular" senses, we can also include feelings of déjà vu, the intense feeling of cold that so many people get in the vicinity of Glencoe in Scotland, where a massacre took place, the immediate feeling of already knowing someone we have just met, the hearing of one's name being called when a loved one is in crisis elsewhere, plus many other examples. These all have in common a quality of vagueness or symbolism rather than the specific quality that ET experiencers report.

It does seem reasonable to suggest that skill in remote viewing and/or telepathy makes sense as a prerequisite for the kind of telepathy that we will need in order to have real conversations with our ET visitors.

While there are undoubtedly people who are born with psychic and/or telepathic abilities, it is sad but true that for almost all of us, development of those skills takes a course of study followed by constant practice. It has in fact been likened to learning one of the martial arts. Hopefully, this will not discourage too many people from making the investment. At least some of the instruction on remote viewing is downloadable for free from the Internet. The Farsight Institute is notable for this and I can attest that the material is well presented and comes in plenty of depth including a printed manual, also downloadable. More formal and intensive training is also available at that and at other institutes but does cost money, and this is fine because not everyone has the self-discipline to go through a course of this type working at home from audio CD's.

In the Ambassador to the Universe training offered by CSETI which I did, it was very noticeable that not everyone saw every instance of an ET signal, although everyone saw at least several. For one thing, we were sitting in a circle, so that in a given case, only about half the people were facing the right direction. Virtually all the signals were very brief in duration, and what was actually there was only able to be captured on a time-lapse photograph. What was clear, was that the flashes of light of different but specific sizes and colors, were seen by multiple people at the same time in the same location, which rules out imagination and tricks of the mind. Most of them were seen low down against a background of trees, bushes, or the sandy beach where we were sitting, which also rules out low flying aircraft, weather balloons, and so on. We were far enough from any buildings or roads that people or cars were also ruled out. There were hotels and beach front property about a mile to the south of us but virtually nothing was seen in that direction, and of course any flashes of light that were noticed from there would be quite rightly ignored. It was noticeable however that many objects, while seen by several people, were not necessarily seen by everyone looking in that direction, which gives credence to the idea that highly developed perceptions are important. Shooting stars were seen, but also short streaks of colored light in the sky that did not look like shooting stars at all.

There is a clandestine feeling to the majority of cases of possible ET contact reported in the literature. Many UFO sightings are only experienced by certain of the people present, while everyone else seems unaware that anything unusual is going on. Again, many UFO sightings are not experienced at the time but come out when photographs that were taken are viewed later on. I have myself had this happen – an orb showed up on a photo taken of a scene at the Mount Washington Hotel in the White Mountains of New Hampshire. (See below). It is not particularly spectacular but there was no dirt on the camera lens as many other photos without this object were taken both shortly before and after this one. Besides, the object has texture and depth (implying focus), unlike the way that dirt particles show up (very fuzzy). I can attest that no object in the air at that time was visible to the eye.

Orb in photo at the Mt. Washington Hotel

This object is clearly unidentified and flying, and there is no way to estimate its size since it could be a few hundred feet away or a couple of miles. It may not even be alien in origin, being perhaps one of our own military's secret surveillance toys. What is really interesting is that whatever its origin, it is confirmation that someone has technology that can conceal their devices from the human eye. The US military has already demonstrated the feasibility of this by having both image sensing and projection on both sides of the object to be camouflaged, and they have demonstrated this both with people and aircraft. The method is to project the view from the other side of the person or aircraft, to the side facing the viewer, making the aircraft or person appear virtually invisible except at the edges. What this means is that these objects are very unlikely to be noticed unless the observer is concentrating their attention on that part of the scenery at the right time. The idea of this was very well done by special effects in the movie "Predator".

The object in the photo appears to be spherical, which would be harder to camouflage by the above technique, because the edge of it would be more clearly defined. This either confirms that one has to be concentrating on the right place in the sky while the object is present, or alternatively that the stealth technology being used is a great deal more sophisticated. The Mount Washington Hotel had plenty of visitors and guests wandering around on that day.

Another experience I myself have had, came to me courtesy of Ed Grimsley mentioned above, who was a presenter at the International UFO Congress early in 2009. He brought to the conference a few military grade night vision binoculars, and each night took a group of people out into the Nevada desert to observe the sky for a couple of hours after dusk. I am still processing what I saw, which was this: there were a number of objects visible up there which were unidentified in that they were obviously not birds, moths, planes, shooting stars or satellites. We saw quite a few of all of those, and their appearance and flight pattern was as one would expect. The objects that were exciting were moving around up there and were changing direction, often rather abruptly, which of course our satellites cannot do except quite gradually by firing off a rocket engine as they change or adjust their orbits. The objects of interest had the same appearance as the fainter stars in that they showed up as small blobs. Also, there was no sign of rocketry when they changed direction. The speeds and directional changes were quite dramatic when you consider that they had to be at least a hundred and probably several hundreds of miles up. An additional point of interest was that a couple of hours after dusk, the frequency of sighting dropped dramatically, which of course places a limit on their altitude at a few hundred miles, since we could only see them in the infrared if they were reflecting some sunlight (or were extremely hot – unlikely since they were in space). Ed Grimsley had been counting the sightings and in total we had seen about 100 objects in the two hours or so. It is hard to understand is how it could be that so many objects were moving around in that limited piece of Nevada sky over just a couple of hours. It seems inconceivable that such a density of activity could be extrapolated all around the world, since that would imply tens of thousands of craft in our near-in space at any time. It is true that southern Nevada is close to several US Air Force bases, namely Edwards, Nellis, and Groome Lake (the infamous Area 51), so one can imagine that some or all of the objects may have been still-secret craft belonging to the US military. But, all of them? Well, since our military has apparently been working with ET technology since at least the Roswell crash in 1947, it certainly seems possible. Another possibility is that a number of ET craft may have congregated in that place just because the premier yearly conference in the area of UFO investigation was happening there and then, and they were observing. All I know is what I saw, and as I say, I am still processing the experience. I would get my own set of those night vision binoculars and recommend that others do the same, except that they sell at around $3000 each. Ouch.

It has now been 64 years since Roswell, and that was by no means the first such event, and certainly not the first sighting. The web of deceit is showing more and more signs of unraveling as increasing numbers of people with the kind of credibility of Edgar Mitchell (Apollo 14 astronaut, scientist, author, and founder of the Institute of Noetic Sciences) come forward. A number of countries (France, Mexico, Russia, China, Spain, Brazil, Canada, India, Turkey) talk openly about UFO activity, and some of those have released most (probably not all, though) of their UFO files. The UK has started to release information too. A large package came out early in 2011.

Any politician would have an understandable reluctance to come right out and say "OK, we have been lying to you for over 60 years, and by the way we have spent many hundreds of billions of your tax dollars on secret programs without acknowledging their existence. Plus, there has been no oversight or accountability whatsoever." Now we know what Donald Rumsfeld was talking about just after taking over the Pentagon, when he said that there was 2.2 trillion dollars he could not account for. That's right, 2.2 TRILLION. So how long do we have to wait for a politician with the honesty and courage to own up to the cover-up? It is going to be a long wait! It is also clear that most of our elected representatives including most if not all of the presidents since Eisenhower, have been kept largely in the dark. Steven Greer in "Hidden Secrets" discloses that at least Kennedy, Ford, Carter and Clinton were aware of the issue. They all tried to get access to the information and get things under control. In all cases they were denied access and in some cases, actually threatened and told to "back off". Greer also states that Hubert Humphrey, vice president under Lyndon Johnson, was concerned about this and was denied access to a restricted area at Sandia Labs in New Mexico when he tried to find out what was going on.

Needless to say this is all deeply disturbing as well as being difficult to believe. Talk about worldviews being threatened! In our (somewhat) safe and comfortable worldview, the President and Vice President are supposed to be at the top of the pecking order and must surely therefore have access to everything that is going on!

It will be quite difficult to persuade the defense industry to reveal and account for the huge amounts of money and alien technology that they have been given over those 60 plus years, and to get them to live happily under a new deal in which that money would now come under the control of Congress and the budgeting process, and in which the ET technology would move into the public domain.

The strategy if there is a conscious one, appears to be to gradually let the information out and throttle the rate of disclosure as much as possible, while keeping the gravy train of "black ops" money chugging along for as long as possible. I should perhaps define what I mean by "black ops" money. It has been well known for a long time that there are public funds that go to agencies and operations that are kept secret, and within limits, that is understandable. I refer more to the $2.2 Trillion that Donald Rumsfeld could not account for when he took over the Pentagon. And that is just the Pentagon! It is, by the way, interesting that not a single member of either house of Congress spoke up about investigating where all that money went. We know of course that there is a division of Lockheed known as the "skunk works" whose chief boasted after his retirement that we are now able to "send ET home". Another direct quote "We already have the means to travel among the stars, but these technologies are locked up in black projects and it would take an Act of God to ever get them out to benefit humanity… anything you can imagine we already know how to do".

It seems to be possible to rely on a huge inertia and apathy on the part of the general public. Recent history has shown that most people's worldview is very deeply entrenched and does not allow the consideration of information which challenges that worldview (such as that our President is not in the loop on issues of global, let alone national, importance). The difficulty of getting acceptance of how we as a race are affecting global warming is another example of this – the almost complete unanimity of the scientific community seems to make very little impression. Yet another example is that it is widely believed that it is better to be screwed by the health insurance industry and the pharmaceutical companies than to put up with some government inefficiency in an equitable healthcare system. And another is the unwillingness of so many people to accept that oil, natural gas and coal are limited resources which are going to run out quite soon, so that it is OK to buy gas-guzzling cars and resist investment in alternative energy.

In many ways the very routine and "normalcy" of our daily lives is a big part of the difficulty. We get up, feed the family, get the kids off to school and/or get ourselves to work, read the paper, get through the work day, and spend our spare time in various forms of relaxation. The seasons go around, the flowers still bloom in the spring, we go to the beach or the lake in the summer, the leaves drop as usual in the fall, and after Thanksgiving and Christmas/Hanukkah/Kwanzaa, winter comes in and does its thing. There seems to be almost no inclination to wonder about what is behind the news and how other people in this world, let alone other worlds, think and live their lives.

Our children are taught but most of them are certainly not educated…

Our children are taught but most of them are certainly not educated – it is appalling to notice how geography, history and current events are not even on the syllabus at most public schools, and this shows for instance, in the scary statistics that show how few children can place well known countries on the map, let alone name their capitals. There are of course some people who are well off enough to be able to send their children to better schools, and who make sure that they know something of the world, and make sure that their families travel. Unfortunately they are in quite a small minority, and especially in the current recession, more and more families are too busy surviving to think beyond the moment. And who can blame them?

It seems always to take a crisis to get some action. Most people are basically very goodhearted and will rally round at once if there is a big storm, a major power outage, or an earthquake. The problem we have here is that there is a whole set of crises headed our way, and while most people will indeed rally round and help each other when each crisis hits, in most cases it will be too late because these crises will be different in that they will be bigger and will last much longer, in fact indefinitely for the most part. When we really do run out of oil, for instance, that is going to be it because it will take many years to do the engineering and put in place the infrastructure for the replacement energy source (even if we are fortunate enough to have a suitable one ready to go) into place. Car pooling and getting around on bicycles or horses is not going to hack it!

The question is, what are we going to do about it? The answer is twofold – first, we need to keep bringing the truth forward, and in as many ways as possible. Second, we need as I said in Chapter 2, to be making our own connections to the ETs that are here and are willing to talk (telepathically or audibly). Given the difficulty to get through to that majority of people with entrenched views, we need to present the truth in different ways, rather than just bludgeoning them with the facts. Without being patronizing, either. This is not a simple thing. I am reminded of a lecture I attended while an undergraduate, which was given at the Cambridge Astronomical Society by the Secretary of the British Flat Earth Society. It seemed like a fun event at the time, but it was actually quite revealing in retrospect. This was a very earnest gentleman whose organization had glommed on to some passages in the Bible that said the Earth was flat, and they took it from there. He used facts like that you can see Calais from Dover, and showed us a map of the world with the North Pole in the center and the South Pole sort of spread around the outside (there was someone in the audience who had been to the South Pole, who was quite amused by this). As the questions started to come in (most of us were so taken aback by his premise that we did not know where to start), he very visibly used data that fitted his ideas, and dismissed as fraudulent or Devil-inspired, data that did not. Because, you see, to him every word in the Bible (that matched his worldview) was absolute truth. The old expression about someone "not wanting to know" is, sadly, literally true in cases like that, and it is also true in the cases of global warming, health care, burgeoning population, and energy. The great teacher in the New Testament paraphrased this when He said "There are none so blind as those who will not see." Logically, it would seem that it should be enough to see Al Gore's movie "An Inconvenient Truth" for everyone to "get it". But in spite of the excellent presentation, the hype, and the awards, it has been largely forgotten. This worldview thing is not susceptible to logic, unfortunately.

It is not enough that the information about ETs, UFOs and the ongoing cover-up is all over the place, in published books as well as the Internet. Unfortunately the very size of the Internet is a problem in that people only have enough time to visit a few sites, and will naturally spend that time on sites that agree with their worldview, if indeed they get beyond Facebook. (In a sad sign of the times, it was just reported that Facebook is now getting more hits than Google, and Twitter cannot be too far behind. Google is a brilliantly executed source of information, whereas Facebook is a means of keeping in touch with friends and family, and 90% of its content is trivial. To be fair, Facebook and Twitter have also been instrumental in helping pro-democracy protests organize themselves around the world, too.) It would make a huge difference if schools encouraged children to spend just a few minutes each day researching some person, event or country on Google or Wikipedia. Both of those sites are great resources, and what a great homework assignment it would be, and not exactly burdensome. Would it not be even greater if the kids could engage their parents in that part of their homework too, and it would be fun for the family as well!

The most promising avenue is to try to get the mainstream media, both newspapers and television channels, to start taking the subject of ET contact seriously. That at least will cause it to come up as a subject of discussion around the dinner table. While the idea of media coverage is promising, though, it has proven to be extremely difficult to obtain. Many clever and scholarly people like Stanton Friedman, Steve Bassett and of course Steven Greer, have been working on it for many years. Books have been written, conferences have been held including press conferences at the National Press Club in Washington (I attended the one in 2010 – it has become a yearly event), numerous web sites have been created, and radio and TV shows have been put on. A list of many of these is at the back of the book.

As an example of media coverage, remember the treatment given to Denis Kucinich during the Presidential primaries in 2007, when he acknowledged his own experience of seeing a UFO? He was skewered! Amazing, considering that both Ford and Carter had experiences of their own, and that most of our presidents since Truman have been well aware that we are being visited.

Chapter 5

Will they respond?

"Universe society does not want us to export war or violence into interstellar or inter-dimensional space." Alfred Lambremont Webre, in "Exopolitics"

Abstract: The answer to this question depends on the agenda that our ETs are working with. In that there seem to be several different groups of them in contact with us at this time, it is probable that they each have their own purpose, or at least that there are several differing purposes. There may be a rather serious reason why it is and probably will continue to be, difficult to get clear and consistent responses. We may be under a form of cosmic quarantine, as implied by the above quote!

Many inquisitive people (including myself) have tried to contact ETs with little or no success. There could be several reasons for this:

1. They are simply not interested – their agenda does not include our participation.

If this is the case, we will just have to wait. However a number of people (including as mentioned in Chapter 1, Whitley Strieber, Lisette Larkins, George LoBuono, Steven Greer, and Jim Sparks and many others), claim to be having conversations fairly routinely.

2. They are interested, but their agenda involves watching our development without interference.

This idea is belied by the fairly constant "interference" which has been so well documented by UFO researchers since the 1940's. While many of the UFO sightings and encounters have the flavor of covert activity by our own government and the obfuscation thereof, a very large number do not.

3. They are only communicating with selected individuals, most of whom for their own reasons are not making this public.

This is quite consistent with what has been reported. A very few courageous people have come forward and have told their stories (like Strieber, Larkins, LoBuono, Greer and Sparks). A number of others have sought help from mental health professionals who specialize in this field, notably the late John Mack. It is reasonable to suppose that there are others who are continuing to interact with ETs but either fear persecution or ridicule, or just prefer to operate in privacy for the time being.

4. As proposed in the last chapter, they are waiting until we are more evolved.

This notion is not really inconsistent with 3. The ETs are willing to talk to the rather small percentage of us that are sufficiently advanced, but are holding back for the rest of us until we stop abusing each other and the planet we are entrusted with.

5. Further than 3, we have been placed in quarantine and both we and the ETs have to wait until it is officially lifted by the galactic powers that be.

This is also not inconsistent with 3 and 4. In this scenario, the quarantine is in place for the Earth as a whole and will remain so. However the process of coming out of it has started with the participation of a small number of people who are ready.

The idea of a cosmic quarantine is quite plausible but hard to verify. If there is one, it would be helpful to know the set of criteria that we have to meet to get out of it! One criterion that has become clear is that we have to abandon nuclear weapons and scalar weapons[13],

---

[13] Scalar weapons derive their destructive power from the Zero Point energy field. In other words they use the same energy source that can (and should) be used for peaceful purposes. Their development is deep inside the black budget.

especially for use in space. There is evidence to show that some of our early experiments in this area have been interfered with, notably the one in which we tried to send a nuclear bomb to the Moon and explode it, and also the unsubtle temporary shutting down of a couple of ICBM sites while UFOs were present. The latter events were testified to by several retired Air Force officers at a news conference at the National Press Club in Washington DC, late in September 2010.

I would suggest that another criterion is that we have to demonstrate that our intentions towards our galactic neighbors are entirely friendly, and that we welcome interaction with them.

In terms of our plan of action, it does not make much difference whether 3, 4, or 5, or some combination of those, is the actual state of affairs. Any one of them will require us to keep searching, to grow in awareness, to solicit contact in any way we can think of, and to network with each other and share ideas and experiences.

It is proper to wonder at this point what the ET reaction to our efforts might be. We have to hope that they have a sense of humor, for one thing. Certainly the present chaos and general nastiness on Earth must reinforce any notion they have that we are not yet ready as a race to enter Galactic society.

First, let us make it a rule that any conclusions we draw must make sense, at least in terms of our current knowledge (meanwhile taking into account that our current knowledge is, so to speak, a moving target). Second, it is also essential to be cautious, while at the same time being ready to accept ideas that are new to us. This is not as inconsistent as it sounds – think of the difference between Science Fiction and Fantasy. A good Sci-Fi book will introduce us to new ideas, but will also be internally consistent and will be based on science that we can accept as possible even if it is not part of our present day paradigm.

However for this analysis, we have to try to put ourselves in the ETs' shoes (whether or not they actually wear shoes, we may soon find out!). We should also in fairness as well as for our own peace of mind, do this exercise for ill-intentioned as well as benevolent ETs. Let us take a deep breath and see what a malevolent ET might be thinking and doing. There seem to be several motivations to consider.

1. Plunder the Earth's resources.
2. Enslave the human race, perhaps as a way of conveniently implementing number 1.
3. Use human DNA as a means of refreshing their depleted and failing gene pool. (This has been reported in a number of books.)
4. Take over the planet so that it can be colonized by the ETs.

The first thought that comes to mind is that none of the above requires stealth, given the very advanced technology to even get here in the first place. If we do put ourselves in their shoes, suppose we were able to get to other inhabited planets and that we had one of the above objectives. If we were uncaring about the feelings and the fate of the native species, what would be the point of pussy-footing around? We would "just do it", wouldn't we?

Secondly, the plundering of the Earth's resources makes sense in some ways, as the resources on any given planet are finite, as we ourselves are presently realizing, so that it is easy to imagine that a race of ETs might have run out of some key elements or minerals in their own solar system. However, even with our limited technology, we already know that there are virtually unlimited resources of all kinds in our own asteroid belt, and an interesting array of minerals and chemicals both organic and inorganic, have just been discovered on our Moon. For someone coming in from elsewhere in space, it would be far easier to work on our asteroids (low gravity, resources close to the surface) than to have to come down our gravity well and then have to cart the materials back out into space again. If we accept Zecharia Sitchin's work translating Sumerian and other ancient records, those records indicate that the Annunaki were active in removing gold from Southern Africa and tin from South America, and did genetically breed humans so as to form a labor force for that mining activity. This anomaly, that they surely could have much more easily extracted what they wanted from our asteroid belt rather than undertaking a lengthy genetic engineering program and then teaching the resultant labor force enough to do the work, will doubtless be resolved at some point. What makes sense to me (based on no evidence), is that if Sitchin's interpretation is correct, the Annunaki at that time were not so very far more advanced than we are at this time, with relatively simple DNA engineering ability, and the asteroids they were able to find were lacking the specific minerals they wanted (gold and tin). Since it seems that our asteroid belt consists of ice balls with iron and/or nickel cores, that is at least plausible. But it does not seem likely that the bulk of the easily accessible gold and tin in our solar system is concentrated on our little planet, which after all formed originally from the same space dust as everything else in our solar system, and would thus have to have roughly the same mix of minerals.

The next thought has to do with item 3 above. It is very hard to believe especially with our own rapidly growing understanding of DNA manipulation, that any highly advanced race of beings would have any trouble making the necessary changes to their DNA if they discovered any signs of deterioration.

Lastly, what about the idea of an ET race overflowing its habitable planets and wanting more living space for its people? Ah, lebensraum. Where have we heard that word before? George LoBuono, in his book "Alien Mind", talks of a race called the Verdants, who are actually engaged in that process in our galaxy as well as others, and even states that there are operatives working for them on Earth right now. This is scary stuff indeed, and hard to verify one way or the other. It is encouraging however, to note that LoBuono does not by any means despair of our ability to resist this operation. It appears that their methods are manipulative rather than directly aggressive, and if enough of us learn to operate in the telepathic/remote viewing world, it will not be difficult for us to hold our own.

It is extremely difficult to wrap one's head around this story of LoBuono's, which presents a cosmic picture of rather overwhelming proportions, involving multiple galaxies and hundreds of millions of years of relevant history. It seems that the Verdants (if we accept what LoBuono is saying) operate more in the role of controlling parents than occupiers. It is disturbing to note that he claims that there are a number of people on Earth in positions of power and influence who are presently working in concert with the Verdants.

Philip Krapf in his book "The Challenge of Contact" also talks of the Verdants, but puts them in a much more benign light. Steven Greer also is adamant that the aliens in contact with us are benign. So now we are presented with a dilemma – what to believe?

It must be realized that there is a phenomenon that crops up with information that is channeled, received via telepathy, or received by remote viewing. That is, the information whatever the source, has necessarily to be processed by the receiver's mind before it can be reported. Thus inevitably, however hard the receiver tries to be objective, there is some filtering going on. Or to put it another way, the information is modified to a greater or lesser extent, by the receiver's worldview! Here we are back with the old worldview issue again!

What this really means is that we have to be cautious with information received mentally, even from people with high integrity. On reflection, we already know that we have to be cautious with information received visually, audibly, or tactually, too! Remember the old saying in legal circles – "no-one lies like an eye witness"? Or another one – "believe nothing you hear and half of what you see"? Taste and smell can be misleading too.

Given the possibility of ill-intentioned ETs as just described, it is not unreasonable to have a certain amount of paranoia! However it needs to be a different kind of paranoia than the one which has our governments working to scare us into spending more and more on our military and its associated defense industries. It is an unsavory truth that an electorate that is running scared will be willing to put up a lot more money for defense than one that feels really secure. We have seen very clearly in recent years how convenient it is to have a collection of bogeymen around to frighten us with, like Osama bin Laden, Mullah Omar, Radko Mladic, and others. If the US intelligence services had as much success with remote viewing as they clearly did have, obviously they know exactly where these people are, and have known for many years! Very recently a couple of those bogeymen have been found – bin Laden and Mladic. Well, it's only been 10 years or so. Meanwhile there is plenty of fuel to keep the pot boiling. Libya, Pakistan, China, North Korea, and Iran, for instance, with several others bubbling away or waiting in the wings, such as Palestine, Syria, Saudi Arabia, Yemen, Sudan, and others.

We do however need many of the qualities of good soldiers – courage, resourcefulness, tenacity, and teamwork.

Paranoia over threats from space will get us into massive spending on the development of space weapons and the craft to deliver them. However, the point is that expensive space weaponry will not be effective against invaders who are not using them, who are already here, and who in any case have no intention of using force on us. The weapons we need are those of the mind, and it is good to know that they are not expensive, nor do they need a large military industrial complex to deploy them. We do however need many of the qualities of good soldiers – courage, resourcefulness, tenacity, and teamwork. It is also good to know that many of the tools we need are being made available by people who were part of the USA's remote viewing program, notably Dale Graff, Russell Targ, Harold

Puthoff, Joe McMoneagle and Lyn Buchanan, and others who were taught by those people, such as Courtney Brown. This program went on in secret for about 23 years before the government discontinued it in 1995. These talented and eloquent people have been publishing and talking about it ever since (a Google search on any of them yields plenty of information).

Turning now to benign or at least neutral ETs, let us look at their possible motivations.

1. Space tourism.
2. A desire to observe our development for research purposes.
3. An ethic that drives them to search out developing races and help them along. A more assertive version of number 2, which would involve working on our DNA.
4. A mentoring role that has them standing ready to work with those of us who are both willing and sufficiently aware.
5. An observational role that is strictly hands off until we reach a certain level of development, at which time we would be welcomed into the galactic community.
6. An observational role that includes putting in place the means to contain us inside our solar system (or even just the Earth and our Moon) or our local part of space so that we will not be able to go out and cause havoc among our space neighbors. This containment would remain in place until we reach a certain level of development. This would explain the effectively zero progress in our space program since the last moon landing almost 40 years ago.

I suppose we should first concede that since there are a number of races apparently here, any and all of the above are possibly in play at this time.

(Perhaps overheard at a top level meeting on the subject – "Yeah, maybe they are nice guys, but they can still make us do things their way, and our freedom is in jeopardy!")

It is important that a central agency like the UN does the collection, characterization and correlation of all the information and disseminates it to everyone, so that we could all benefit from the new technology. It would also be nice for us all to have some idea what to expect when we come into contact with a particular group of ETs.

The temptation for the powers that be to withhold or distort the information, particularly if there is perceived value because of new technology involved, is just too great. In fact, there seems to be plenty of evidence that the US military/industrial complex has been doing just that – withholding technology that ETs have shared and technology that has been reverse-engineered from crashed ET craft - for many decades now. What we need to do is find a way to remove the profit motive, and also the hoary old excuse that national security is involved. It may well be necessary to go public with some of the technologies without waiting for disclosure. Those parts of the military-industrial world that have been working with those technologies for all this time will then be embarrassed into disclosure, hopefully. Maybe.

There are definitely things we can do, however, and we can do them now.

That change of modus operandi is a lot easier to talk about than it is to make it happen. There are definitely things we can do, however, and we can do them now.

- Those of us who are willing need to develop skill at telepathy and remote viewing. There are plenty of courses on both of these disciplines (and also on channeling, incidentally) on the Internet and at various institutes around the USA and the world. A special case of this that is directly applicable, is the Ambassador Trainings offered by CSETI. These are not for everyone, though, and require a serious amount of discipline and commitment. They are definitely not for the simply curious.
- Study the field in depth. There is quite a good bibliography at the end of this book.
- Go to conferences and develop networks and support systems.
- Become more observant of the sky and of the environment. It does seem that there are opportunities to see and experience UFOs and ETs. We have to learn to notice anomalous sights and sounds, and be willing to see and hear things other than what we would normally expect. All too often we see and hear what we want and expect to see and hear.
- Form our own opinions on what is genuine ET activity as opposed to the military playing with its advanced toys, and be critical and analytical of any actual sightings or encounters.

Chapter 6

The Military Mind and Threatened Worldviews

"Human beings have an astonishing ability to dismiss information that does not conform to their preconceived notions of reality." Courtney Brown, in "Cosmic Voyage"

Abstract: As discussed in the last chapter, the behavior of the military since World War II is not that hard to understand. Towards the end of that conflict, "Foo fighters" were toying with both Allied and German aircraft, and both sides realized that they were dealing with craft of unknown origin. Since then, there has been a veil of secrecy over the whole subject of the ET presence which remains in place even today. This means that we have to rely on a few brave souls who have been involved in this, to come forward and tell what they know.

Before we go further with this discussion, I realize that many readers will still be struggling with a gut-level resistance to the information and ideas being presented. While some will have looked carefully at the body of evidence and decided to reject it, to many the subject is interesting, but new and troubling.

This reaction goes beyond mere doubt, because the presence of extra-terrestrial beings here and now is directly threatening to most people's understanding of the world (their worldview), especially if they hold strong religious beliefs. In John Mack's book "Passport to the Cosmos", he wrote that our worldview holds our collective psyche together, and that society will find ways to punish anyone that puts forward ideas that challenge or undermine the mainstream worldview. Later on in the same chapter of "Cosmic Voyage" that was quoted in the chapter heading, Courtney Brown goes on to say:

> "Externally obtained information can come from a newspaper, a friend, a lecturer in a university, a book, or any other source. But when confronted with ideas, let alone facts, that do not fit into an accepted informational paradigm, humans tend to have an intense desire *not* to believe the new information. At times it seems any excuse will seem rational, since the goal is what matters: the established paradigm must not be readily abandoned."

Examples of people who have been punished severely because of that reaction include:

- John Mack himself, who was subjected to an intense campaign to take away his tenure at Harvard, for his contention that ET experiencers were not hallucinating and had indeed had extraordinary experiences.
- Wilhelm Reich, who died in prison for his experimental (and effective) work on alternative healing and new forms of energy.
- Galileo, for proclaiming that the Earth was not the center of the Universe.
- Giordano Bruno, who was burned at the stake for proposing an infinite universe with a unitary Creator.
- Nicola Tesla, certainly one of the most brilliant minds of the last 100 years, saw one invention of his after another rejected or taken over by others. Only recently has the full extent of his work started to emerge.

There are many stories of inventors who had their laboratories mysteriously destroyed, and/or their work sabotaged. There are far too many of these to be dismissed as coincidence. The point here is, what is useful about destroying the work and/or reputation of a harmless crackpot inventor whose work is going nowhere? There is however plenty of point in doing so to an inventor whose work if brought to fruition, will threaten the profits and/or assets of one or more major corporations! To get some perspective on this, consider that the total assets of the fossil fuel industry (that is, what remains in the ground, plus all the plant and infrastructure) have been estimated at about $600 TRILLION. It is not hard to believe that the industry is willing to take extreme measures to prevent any inexpensive and non-polluting source of energy from seeing the light of day.

Let us try to find our way to some conclusions then, and see if they make sense. It is perhaps helpful to start by putting ourselves in the position of the primary actors in the fascinating drama that we are all a part of (which sadly has regular descent into farce and/or tragedy, as we can see daily by reading the papers).

Perhaps the most important actors on our side are the various governments. Whether we like it or not, it is their primary job to keep us safe, and this necessarily calls for a certain degree of paranoia. It is not necessary to believe in a sinister elite group of shadowy powerful people who manipulate all the important stuff behind the scenes (although there is in fact a lot of evidence that such a group does exist) to accept that much. So if we imagine even a responsible and upright government, and then expose them to not one but a number of races of ET visitors, what do we imagine their response will be? Also throw in that these visitors come in craft that can run rings around our best aircraft and dance in and out of radar as well as visual contact, and remember that their technology has allowed them to easily cross interstellar space. It is really not surprising that our governments react with a lot of concern! While it is possible that all these visitors are benign, does our military have the right to make that assumption, given that their job is to protect us? OK, let's be real. Of course not. So, what would we do in their place?

- Shoot the ETs down? One would think not – after all, with the technology that they must have, the ETs could incinerate us all without breaking a sweat. But in fact there are reports that the military has tried to shoot down UFOs a number of times in the last 70 years. We should be very relieved that the ETs involved either bobbed and weaved, or in some cases just destroyed our aircraft (there is evidence that both the USA and the Soviets lost many planes that way during the 50's and 60's), rather than exact a painful vengeance on all of us.

- Try to talk? Definitely, but suppose the ETs do not want to talk, or that they do not come up with enough information to satisfy the fears of the military?

The most likely reaction is what we appear to have seen over the last 60 years or so. A very high priority program to investigate the phenomenon, in complete secrecy as far as the general public is concerned, is really the only option, especially in the context of the Cold War atmosphere of the 1940's and 50's when these decisions were taken. When some ET craft crashes because of equipment malfunction (it is hard to believe that even the most advanced technology does not go wrong sometimes), grab the ETs and their vehicle and analyze the bejesus out of them.

Now the whole scenario proposed by Michael Mannion in "Mindshift" makes sense, including the regular muddying of the waters by our governments. As noted earlier, many of the reported incidents including (and perhaps especially) the abductions, animal mutilations and surveillance drones (silvery orbs, "dragonfly" machines, etc.), have the flavor of our own government testing its new technologies from whatever source. Those types of incident do not at all have the flavor of actions of ETs with vastly superior technologies, whether benign or not. Why would they need to do any of those things? It does not make sense, and we should recall rule No 1 above, that everything does need to make sense in terms of our knowledge at any given time, knowing that as we learn more, what makes sense may well change.

It is also true that the events surrounding the Roswell incident make all kinds of sense too, when looked at from this viewpoint. An ET craft crashed, the USA military grabbed it, perhaps with one or more occupants still alive, and have been working to understand and replicate the technologies ever since. They have apparently met with some success, if we buy that the abductions, animal mutilations and surveillance drones are evidence of that activity.

What is upsetting and depressing about all this is that it promotes the idea of scary, malevolent and unfeeling ETs. It is also sad that our government has done nothing to dispel the idea, allowing the image of the sinister ET to proliferate. The conclusion of Steven Greer and others is that the growth of power and riches in the military-industrial complex has led to a self-sustaining hubris and arrogance which is served well by the idea of sinister ETs.

It is of course much easier to imagine the behavior and motivation of our governments than it is that of the ETs. Our governments are, after all, composed of human beings with all the attendant paranoias and psychoses. We have plenty of examples of paranoia, corruption, general craziness and disastrous errors of judgment, not only in the last administration (George W. Bush, 2000 – 2008) but in every one in recent memory to a greater or lesser extent.

Unfortunately, "National Security" can be, and is, used to justify almost anything, from water boarding to murder. However in this case the security that is being protected is not that of the USA or even the world. It is the security of the status quo, which sustains and brings wealth and power to a very small number of people, while leaving the rest of us in various states of struggle and ignorance. Also unfortunately, when the National Security card is played, any questioning of its validity is characterized as unpatriotic and as jeopardizing the safety of "our mothers and children". And that is so effective at suppressing dissent! Even in the unlikely event of the people who play that game being brought to justice, the plea of doing what they did in the cause of National Security will be very effective.

It is not of course only in the USA and the West that this mindset is in vogue. Essentially the same tactic is used even more blatantly in China, Russia, Iran and Burma just to name a few. In China for instance, dissent is suppressed quite violently and publicly in the name of "stability and harmony" (remember Tiananmen Square?), and the Dalai Lama of all people described as a subversive, a "splittist", and worse.

This worldwide resistance to change in the status quo is still solidly in place when it comes to the ET presence. There is, especially in the USA, an attitude in elite circles that says "we cannot let the cat out of the bag because the people could not handle it". This kind of arrogant condescension has been around at least since the time of the Roman Empire. They referred to the general population as "the plebs". More recently, the French aristocracy (and the revolutionary elite that succeeded them) called them "la canaille". In England they were "the rabble" or "the riff-raff", "the mob" or even "the peasants". In Imperial Russia, they were "the moujiks", and in Soviet Russia they were "the proletariat". In every case these derogatory epithets are about the assumption that the elite know what is best for the people, and those ordinary people had better just suck it up or take the consequences. And heaven help anyone who gets uppity and questions the way things work.

One excuse that is frequently given for the attitude that "the people could not handle it", is the aftermath of the famous War of the Worlds radio broadcast in 1938 where the announcer (Orson Welles) in the story says that the Martians have landed. Many people apparently panicked even though the radio channel had issued warnings that it was just a story. In response to that, it is now over 70 years later, and it is noticeable that in the meantime the public has not freaked out over similar incidents in living color in for instance Star Trek, Star Wars, Alien, Independence Day, or the Visitors.

It really is extraordinary that this mindset of treating the public as children still endures. Or is it? Perhaps it endures because it is a convenient excuse to keep these secrets closely held. It is certainly a well known fact that people of all ages and backgrounds are liable to behave like children if they are treated as such. The converse, that people will behave in an adult manner when treated as adults, is equally well known but much less frequently recognized and acted on.

The military mind, of course, represents a worldview all of its own. It is trained to look for threats, to make detailed plans for responding to different kinds of attack from any possible enemy, and also to make plans for the possibility of making pre-emptive attacks of its own. It has to think about the possibility of allies changing sides, and to gather as much intelligence as possible, not only about known enemies and potential enemies, but also about neutrals and allies. The military has to train incessantly for warfare under all conceivable conditions of terrain and weather, to develop ever more sophisticated weapons, and to play rapid catch-up when intelligence indicates that a real or imagined enemy has superior weaponry of any kind. It is also extremely reluctant to let go of any weapons that have been developed, however horrible. One has only to look at the tenacious grip of the military (of all those countries that have them), on nuclear weapons, biological weapons, and of course mines, and the frantic efforts of many countries that do not have certain weapons, to get hold of them or figure out how to make them. Apart from the USA, UK, France, Russia and China who have had nuclear weapons since the 1960's or earlier, India, Pakistan, North Korea and Israel have already developed nuclear weapons, we know that Iraq under Saddam Hussein tried hard, that Syria took a shot at it a very short while ago, and it is clear that Iran is well along the way.

The worldview of the terrorist is even more scary, of course. Because terrorists are fighting against superior odds and superior weaponry, they have to employ desperate measures such as suicide attacks, and they have to improvise weapons with minimal resources. It is amazing and depressing to note how successful they have been. It is even more depressing to realize that just because they do not have superior weapons, that does not mean that they are not trying very hard to buy or steal them, and that they would not hesitate to use them if they did get hold of them.

The terrorist worldview is one that comes from a combination of desperation and indoctrination. It is very hard to forgive those teachers of extremist Islam that convince idealistic young people that they are assured of paradise if they don explosive vests and detonate them in a crowded marketplace. Many Muslim women actually believe in and try to justify a life that has them veiled, confined to their houses, denied education, and forbidden to go out without a male relative accompanying them.

There are other worldviews that are hard to understand, but are nevertheless real. That preacher in the Southern US that wanted to burn Korans, for instance. Hitler and his gang had a worldview of their innate supremacy – they were the Herrenvolk, and sincerely shouted "Gott mit uns" (God with us) at their rallies. And it was not just Hitler and his immediate cronies, sadly millions of Germans agreed. Those rallies were well attended. This is not to single out the Germans, I would add, the same kind of thing was going on in many countries at that time, and there are some clear signs of fascism on the political right in the US even today. The Soviets had a worldview in which they and their ideology were superior to everyone else's and would therefore prevail.

Many people, notably the Amish, have a worldview that distrusts technology. On a lighter note, the French, bless them, are presently rioting in the streets at the idea that they might have to retire at 62 instead of 60! Meanwhile here we are in the USA wedded to the idea of retiring later and later – the official age is already over 66.

All these examples are by way of emphasizing that there is a wide spectrum of worldviews out there, and I could quote many more of them.

A worldview that now has to be let go of, is the very widely held one that the Earth is unique and is the only bearer of intelligent life. As planets numbering in the billions are now being discovered that are in the regions around their parent stars that have liquid water, more people will be coming round to the idea that there must be primitive organisms elsewhere in our galaxy. But we are still a long way from general acceptance of a multitude of other intelligent races around us, the vast majority of whom are greatly ahead of us in their development.

Chapter 7

What questions should we be asking?

And, how will we know we are getting straight answers?

I believe the questions fall into five categories:

- Science and technology
- History
- Philosophy
- Policy
- The Current Situation

Science and technology go without saying, and history will help us get perspective. But philosophy is important because we must understand what the various ET races have for a set of ethics and how their ideas are shaping their interactions with us and the other races in our cosmic neighborhood. Otherwise we will not have information that will enable us as a race to grow in stature enough to be worthy to sit at that "great table in the sky". Policy and the current situation will give us an understanding of what is going on now and how it is affecting and will continue to affect us.

Obviously other subjects are of interest, such as art and music. But I think we need to focus on the basics for now, at least until we have become junior members of galactic society!

Ultimately the answers to these questions will give us direction in terms of what we can and should be doing to move events in a positive way.

In the case of the scientific questions, they will be verifiable. With the philosophical ones, the agenda of the responders and their own beliefs will inevitably color the answers that we get. With the historical ones, some will be verifiable, some will not. We should also be aware that the beings we will be talking to will not necessarily be representatives at the highest level in their own world. They may well be academics, or tourists! Or, perhaps, opportunistic entrepreneurs looking to make the equivalent of a "fast buck" in their world.

Certainly that information which we get in answer to our questions that is verifiable should be disseminated and verified at once, preferably by several institutions working independently of each other. It is vital that this information not be restricted or classified, so that as many people as possible that are both capable and interested, can do their own verification. Of course, the results of the verification should also be widely disseminated. Fortunately when dialogue with our ET neighbors is commonplace, the nature of the situation will make it very difficult for governmental and military organizations to control the dialogue and the spread of information.

The philosophical information should be dealt with to some extent at the governmental level in that a well thought out and coordinated approach will be called for going forward. Whether such clear thinking and coordination will in fact occur is another matter. One danger will be that military paranoia will take over. Part of our advance planning then, should be to internationalize and spread the responsibility very widely from the beginning. Academia, industry, think tanks, philosophers and yes, science fiction authors (who have done way more creative thinking on this than anyone else), should be involved.

Policy is vital because we have to understand how the specific relationship between us and our ET neighbors is being handled. Only then can we formulate our own policies.

Current events really speak for themselves, are inherently verifiable, and give us a context in which to apply our policies.

Let us make preliminary lists of useful questions, so as to have a starting point. Obviously everyone will have their own thoughts and priorities, but I think it is important to have some ideas before starting. After all, we do not know how patient these beings will be if we start in by asking frivolous questions. My hope is that these lists will serve as a useful starting point.

Science and Technology

1. Are you willing to help us with our energy needs, not only with new technologies but the means to make them generally available before it is too late, that is, before we run out of fossil fuels?
2. Are you willing to help us with our food and drinking water shortages, again before our problems reach a crisis point?
3. Is it actually possible to feed and raise the standard of living of our expected population of 10 billion or more with the technologies that you have available?
4. What is the limit of population size for our planet with the capabilities that you have?
5. What is your definition of the limit of population size in terms of a reasonable standard of living?
6. Do you have methods of healing the damage that we have done to our environment?
7. In particular, do you have a method of rendering radioactive waste harmless without waiting for its natural decay?
8. Do you have a method of population control that can be implemented without coercion?
9. Are you willing to share with us the methods you use to travel in space?

10. Is there an anti-gravity technology that you are willing to share with us?
11. Are there methods of accelerated learning that you are willing to share with us?
12. Are you willing to share with us your methods of manipulating DNA?
13. Do you have the same sense of time as we do? If not, what are we missing in our scientific paradigm?
14. Please correct any misconceptions we may have about the speed of light, recession of distant galaxies, red shift, dark energy and matter, the nature of black holes, the nature of time, the big bang, etc.
15. Please correct any misconceptions we may have about the nature of the atom, sub-atomic particles, wave-particle duality, string theory, etc.
16. Do any of the technologies that you have for anti-gravity, faster than light travel, and energy extraction from the zero point field, have a down side, and if so what?
17. What can we do collectively and as individuals, to make it easier for you to communicate with us?
18. Is there some threshold of capability or talent that we have to attain to make contact with you, and if so, how do you define and quantify that threshold?

History

1. When did your race first visit us here? And how and where can we find signs of this?
2. Was your race involved in manipulating our DNA? If so, what was your starting point and how did you do it? Was it a multi-step process?
3. Has your race had a base here continually since your first visit, or were there visits every so often?

4. Has your race had relationships with previous governments? If so, when and with which, and what were the agreements that were arrived at?
5. Has your race ever interfered directly with wars and other conflicts in our history? If so, which wars and what was the nature of the interference? And what evidence can we find of this?
6. Are there specific well-known historical people that were influenced by your race (other than the governments asked about above)?
7. Was your race involved with any of the well-known UFO sightings? If so, which?
8. Was your race involved with any of the well-known UFO crashes, such as the Roswell one? If so, what was the cause of the crash?
9. Which other races have been involved with us over the millennia, and how far back?
10. How would you describe the advances in your own technologies since you first visited the Earth?
11. Please tell us something of your own history, with emphasis on how you came through the various crises in your development. We are particularly interested in any crisis which is similar in nature to the situation on our planet at this time.

Philosophy

1. How would you describe your race's attitude to humans? How much does this vary among your individual people?
2. Is there an over-arching legal system in the galaxy, and if so, how does it work? Do all the races that are involved with us abide by it?
3. Are there any what we would think of as criminal or hostile elements in the galaxy that have been here or that we may come across?

4. Please tell us your understanding of reincarnation and karma.
5. If reincarnation is anything like the human teachers of it describe, does it take place exclusively within each race of beings, or do we experience lives in many different races? If we have misconceptions about reincarnation and how it works, please correct them.
6. Please correct any misconceptions we may have about the nature of the soul and its relationship to our earthly bodies.

Policy

1. Are we in fact in a form of quarantine? If so, how does it work and what are the set of criteria that we have to meet in order to emerge from it?
2. Clearly the fact that certain people have been in contact for some time, and that there has been some interference with the development of space weaponry and ICBM sites, indicates that the quarantine is neither complete nor simple. How do you decide when to act, and how do you decide whom to interact with?
3. Are there cooperative arrangements between you and any Earth governments, corporations or institutions? If so, what are they?
4. Are there any hostile relationships between yourselves and any Earth governments? If so, which governments, and are any of those hostilities active? If so, are your people and their vehicles in actual danger, or can you evade any hostile actions?
5. What are your best and highest hopes for your relationship with the people on our planet?
6. How would you describe the evolution of your policies towards us since you first visited the Earth?

Current Events

1. What is the status of your presence in our solar system? Do you have bases/outposts on the Earth? On our moon? On Mars? Elsewhere?
2. How many individuals of your race are presently in our solar system? How many vehicles?
3. How many individuals of your race are actively available to communicate with us?
4. Given that there seem to be many ET races visiting us at this time, how would you describe your relationships with the other races, and to what extent is there a community of purpose among you?
5. We would like to understand the thinking and technology behind crop formations. Who is creating them and what is their purpose?

Chapter 7

What can we look forward to, short term?

"Whatever Nature has in store for Mankind, unpleasant as it may be, Men must accept, for Ignorance is never better than Knowledge." Enrico Fermi

Abstract: In what to expect, there will of course be a wide gap between a near future which results from an open minded embrace of the possibilities, and one which results from carrying on with the same way of doing things, the way of being and doing that has got us into so much trouble. This book clearly advocates the former, and there are many leaps forward in technology, agriculture and medicine to which we can look forward. But only if the information that we get is widely disseminated and treated as a set of tools that let us bring in a set of universal benefits.

Something that has held us back for a long time in this process and continues to do so, is the compulsive desire by those in positions of influence who are involved, to preserve their positions and to use the situation to increase their personal fortunes. This is understandable although it is undesirable, unethical and in many cases criminal. This is the paradigm of greed that brought us for example Enron, the housing bubble and its accompanying financial crisis, and two draining and unwinnable occupations[14] in Iraq and Afghanistan.

---

[14] I refuse to call them wars. The war in Afghanistan was over in a matter of weeks and is now an occupation. Likewise the war in Iraq. To that extent, George W. Bush was right when he said "Mission Accomplished". Unfortunately the subsequent occupation was, and continues to be, botched.

Certainly there have been and still are, opportunities to benefit personally from the powerful and exciting new technologies that will become available, but we have also to understand that there is a different choice, which is to embrace the new technologies while moving into a new paradigm of more general abundance and fulfillment.

Let us look at how these two paradigms if followed, are likely to play out and also see if there is any middle ground.

The choice of the new positive paradigm will move us into a world that looks and feels markedly different from our current one. Our present economic system both assumes and requires continual inflation, continual growth in population and in consumption, and a financial climate which is focused on "the fast buck", otherwise known as greed. It is a pity that we have somehow forgotten that capitalism is defined as "enlightened self-interest". The "fast buck" certainly leaves out the enlightened part.

First, let us look at what going with the positive paradigm will mean. Of necessity we need a new worldview. However powerful the new technologies will be, our planet and its resources are still finite. Somewhere there is a limit to the population that we can comfortably support here, and arguably we passed that number of people some time ago. Not only oil, coal and natural gas are limited and being rapidly exhausted, but so also are vital metals, notably copper, gold, platinum, palladium, manganese, titanium and some important rare earth metals such as neodymium and samarium. One also wonders about noble gases like helium, krypton, and neon. (Argon is quite plentiful in the atmosphere.)

It is a pity that most people do not realize that fertilizers, plastics, paint, and artificial fibers such as nylon and rayon are all derived from oil or coal, and all of them are lost in one way or another after use. Except that it gets worse, as some of them do damage in the process of passing into the environment. Fertilizers end up in aquifers or rivers and deteriorate water supplies and fisheries. All too much of our plastic ends up in our oceans and forms ugly islands like the one in the northern Pacific, or becomes traps for fish and dolphins. Paint of course is toxic waste and has to be disposed of carefully. In my experience as a homeowner, it seems that we end up throwing out almost as much paint as we use, as most paint cans are eventually disposed of with a good fraction of their (dried out and unusable) contents still inside.

The positive paradigm is based on sustainability and balance. This means no fast buck, because it means taking a long view of the consequences of our actions and investments. The legalized gambling casino that is the stock market, where the important thing is this quarter's (or even this month's) results, will have to become a legitimate investing arena where the total cost of doing business is taken into account, and where a company's plans for the next decade and more, are at least as important as the quarterly profits. In fact it is hard to imagine a sustainable world where there is such a thing as short term investing. After all, investing was never meant to be about shuffling electronic money in cyberspace, with ephemeral devices such as hedge funds, derivatives, arbitrage, and puts and pops. Currency used to be, and still should be, based on some actual value rather than by a fiat from some faceless entity such as the IMF or the Federal Reserve. It would make sense if that value rather than being based on a single substance like gold, was based on a combination of things so that if any component of it changed significantly in value, the overall value would change only slightly. It must no longer be possible for a polluting corporation when caught in the act of ruining their environment, to simply declare bankruptcy and have the perpetrating executives walk away without being held accountable.

Transforming the general attitude to money will not be easy. We are conditioned almost from birth to believe that hard work and talent will give us a chance to "make it". And it is true that a very small number of people do actually "make it" and become millionaires or more. This is also known as "the American Dream". In spite of the tiny percentage that realize that dream, the majority of us still cling to it. The massive resistance to the idea of an inheritance tax is an example of this. Even though it only affects a very small percentage of the population, those with a net worth in the many millions, most people think that they might still get there and that their family would therefore be short changed when they die. The new worldview needs to be one where we embrace the concept of "enough", and that it is not useful or healthy to be constantly pursuing material wealth.

It may be quite hard to prevent short term speculation from happening. After all, stock markets exist in just about every country, and making that form of gambling illegal is going to be at least as hard as making any other form of gambling illegal.

Another very important way that the new paradigm can be based on reality has to do with our attitude to energy. It was shown some time ago in Dr. Colin Campbell's book "The Coming Oil Crisis" that if the true costs of oil extraction, processing and transportation are taken into account, plus all the subsidies, military protection and other tax breaks and credits, the actual cost to us per gallon of gasoline is between $6 and $15, depending on how you treat the cost of remediation of the damage done to the environment during the extraction and transportation of oil. Those extra costs are hidden from us because we do not see them at the pump – they are taken in the form of taxes, and also remediation of the environment is almost non-existent. (Visit the oil fields in the Niger Delta, the Amazon rain forest, the Canadian tar sands, and the Alaska North Slope if you need to confirm this.) The artificially low price of gas at the pump is a form of shell game; it encourages us to believe that we can afford to drive large gas guzzling trucks and SUVs. It is quite amusing now to remember when Bill Clinton proposed putting a new tax of 6 cents or so on gas (it was just over $2 per gallon at the time) to help pay for development of clean energy. There were howls of protest about how the trucking industry would go down the tubes and the economy would crash. Gas has been well over $3 in the last few years, and has just exceeded that number again, in fact in some parts of the country at the time of writing, it is over $4. Nobody is blaming the recession on that, and the trucking industry is doing just fine. So there is a rather obvious step to take here, which is to take the actual cost of gas and apply it at the pump, while making it very plain where that cost is coming from.

The proportion of the Pentagon's budget that protects our sources of oil and its transportation should be identified, and the total of the subsidies and tax breaks that are given to the oil industry.

The addition of these costs to the cost at the pump should of course be accompanied by an equivalent reduction in other taxes, largely the income tax.

In the same way, the true cost of coal and natural gas should be reflected in what we pay for electricity and fuel. It has recently come to light that, far from being the benign source of energy that it purports to be, natural gas is quite harmful in that the process of extraction involves fracturing of the rocks that contain it, (known in the trade as "fracking"). This results in serious contamination of the water supplies that are local to the gas well. Serious, that is, if you consider the ability to flame up ones kitchen faucet as well as not being able to drink what eventually comes out, a serious matter. Remediation of this kind of damage would be extremely expensive if it is even possible.

The practice of lopping off the tops of the hills of West Virginia and elsewhere as part of the coal mining process is also ugly and causes major pollution of lakes and streams in the local areas. Proper remediation of that kind of damage would of course make the extraction of the coal by those means uneconomical.

In a recent study by Harvard Medical School, it was found that the true *annual* cost of coal mining in the USA is between $350 billion and $500 billion, making the actual cost of coal-generated power per kilowatt-hour much more than that generated by wind or solar technology. Google "Harvard Coal Study" for the full story on this.

This kind of study does not even consider how we are going to pay for the development of new and sustainable sources of energy. Although if the true cost of the way we are doing things was made clear, it would certainly focus the public's attention on the urgent need to find those funds. Those funds by the way, in all fairness should also include money to retrain those workers whose livelihoods have been in the extraction of fossil fuels of whatever kind. The situation is not their fault.

It ought to be obvious that the new energy sources, their manufacture, installation and maintenance, and the building of infrastructures to support them, will be a huge source of jobs, that are a whole lot less dangerous and unhealthy than the ones they will replace. Once the fossil fuel energy workers get that, one would hope that they would embrace the change and the new opportunities that will ensue.

A much more serious approach to recycling will also be a part of a sustainable paradigm. A recycling plan really needs to be a part of all product design. It is actually quite painful to notice how nowadays it is next to impossible to recycle such things as electronic equipment, automobiles, lawn furniture, appliances both large and small, and house furniture. It seems that only a minority of us are doing a reasonable job of recycling paper, aluminum and tin cans, plastics, and glass. It would be a shame to have to make serious recycling mandatory, as some well-intentioned but wrong-headed cities have been trying to do lately. Other towns and cities have taken an equally stupid approach by requiring the public to pay for recycling instead of providing them with incentives to do so. That just irritates people and makes them more likely to quietly dump their recyclables in the woods. The answer is to approach recycling with this new worldview, and then it will come naturally. Education will be a large part of this, since most people simply do not understand the issues. Once the new paradigm has been accepted, there will be a willingness, even a hunger, for that education.

I hope that the reader is starting to see a pattern here, one that shows that the new paradigm is actually based on reality, not some idealistic dream! It will take leadership at the top and adaptability by the working population to make the transition possible in the various industries – transportation, energy, and of course the "military/industrial complex".

The world economy and ecology do not at first glance have anything to do with the ET presence here, but I hope that it will become clear that it is all closely linked and will become more so. Ponder this: if one imagines a civilization that is hundreds of thousands or even millions of years ahead of us in development, it becomes obvious that they have to have solved this worldview/paradigm issue, and the related issues of energy, raw material depletion, recycling, population, and waste disposal. Part of the help they can give us is to educate us on how that worked for them and what the transition looked like. Of the millions of races in this galaxy alone that must have already gone through their version of this crisis of ours, it is obvious that some races will have destroyed themselves, some will have come through with a lot of struggle and pain, and some will have "got the message" early on and emerged with a minimum of pain. The ETs will be able to tell us also if this stage of our development is a cusp, where there is in fact no middle ground, forcing us into a clear decision, or if there is a spectrum of choices and outcomes. Most people who write on the subject position our situation as a cusp, and that may well be the case. But it would be really good to know what other races have gone through.

Col. Philip Corso's book, "The Day After Roswell", spelled out clearly how technologies from the Roswell craft (fiber optics, lasers and semiconductors among others) found their way into defense-related industries. Semiconductors alone have dramatically transformed our lives and our economies since then, lasers are involved in everything from carpentry to eye surgery to communications to CD and DVD players, and fiber optics are deeply involved in the medical field and also of course in communications. Other more exotic technologies have disappeared into the black-budgeted arena. When these technologies come into the public domain, and the sooner that happens the better, several forces will come into play, and it will help a lot if we can foresee them and do some advance planning.

The first force will be anger. It cannot exactly be claimed that all the information about the ET presence and the technologies that they bring with them is totally secret. After all, Corso's book and so many others make it quite clear what has been going on, and there is evidence all over the place if one looks for it. The problem has been and remains, how to get the general public to take notice. One gets the feeling of pressure building, and the explosion will be more violent the longer the lid is kept on. Something that gets people very steamed up, to continue the metaphor, is the realization that they have been cheated and lied to. And that is indeed what the people will realize, once they get that it was their tax dollars that paid for the reverse engineering and the exploitation of those technologies, and which made it possible for a small number of beneficiaries to get very rich indeed.

Since anger, even and perhaps especially righteous anger, can be very destructive, it is extremely important to add elements of understanding and forgiveness to our new worldview. We need a recognition that anger and the desire for revenge do not serve us, but rather divert our energy away from the real goal, which is a world in balance. We need in short to find the way that Nelson Mandela found in South Africa, the way that the Dalai Lama advocates for Tibet, and the way that Aung San Suu Kyi advocates for Burma; that is, the way of "truth and reconciliation". We need to remember the excesses of the French and Russian revolutions and how long it took those societies to recover and move forward. It took the French many decades, and arguably the Russians are still in the recovery process nearly a full century later. We must not let our anger, however justifiable, drag us into that quagmire.

Part of defusing the tendency to violence will be to ensure that those who have been complicit in keeping these technologies secret, do not in any way profit from their complicity, and this should not only be done, but be clearly seen to be done. Any patents that have been granted to people or companies as part of the decades of operations described in Corso's book "The Day After Roswell", should be voided and the information placed in the public domain. Those that have resorted to violence to keep their secrets will have to be held accountable, and will need to make restitution to their victims and their families.

The second force that will come into play is, sadly, greed. Just about every technology corporation (that has not already been privy to these secrets), will want to grab as much and as many of the new technologies as they can, and frantically try to patent as many adaptations as possible of them for their own exclusive use. It is vital that we not go that way, as it will drag us back into the wasteful, greed-based paradigm that we are trying to find our way out of.

We need to find ways of compensating and rewarding hard work and innovation without restricting the flow of knowledge. As an example of that possibility, I am reminded of the early days of semiconductor research (in which I was involved), when the people doing the research talked quite freely, the various conferences were very dynamic, and rapid progress was made by everyone as a result. The papers given at these conferences were full of substance and were delivered by the people who had done the work. But as time went on, more and more information became "proprietary" and was withheld. The papers were delivered by marketing types instead of engineers and scientists, and were vehicles to show how clever their companies were instead of sharing information that had substance.

It may well be necessary to do away with the awarding of patents, or at least to dramatically change the way that process works. We have seen especially in recent years, how the patenting process has been used to bury promising inventions in the name of, you guessed it, "National Security"[15]. The medical field especially has distorted and misused the process for their own purposes. There have been and still are ongoing, attempts to patent genetically engineered plants

---

[15] There is in fact existing law, embodied in the Patent Application Act of 1995, which severely limits and delays the awarding of patents by allowing corporations to challenge the application without even identifying themselves. For energy-related inventions, it is even worse, if the invention "appears to endanger the oil, coal, or gas industry". Under Section 181 of US Patent Law, not only the Patent Office, but also the Atomic Energy Commission, the Secretary of Defense, and the chief officer of any other department of the government designated by the President as a defense agency, have the power to "withhold the grant of a patent". The mind boggles.

and animals, for instance. Very repugnant. There has to be a better way to make sure that the inventor gets recognition and a reasonable reward, without risking losing his or her work, and without the idea being so tied up that no-one else can work with it for 17 years. It should certainly not be possible for a corporation or individual to buy and bury a technology that threatens their business. Unfortunately, like short term speculative investing, the ability to quietly buy and bury a technology will be very difficult to entirely suppress.

Now on to what continuing with the current paradigm will look like over the next few decades.

As key resources like oil, drinking water, fisheries, and arable land become more and more scarce, we can without doubt expect conflict, and in fact we already have it. Much of the fighting in the Middle East has been over oil. The USA has western states arguing over who "owns" the water in the Rio Grande, Colorado, and other rivers. Millions of people worldwide do not even have decent water supplies to argue over. Fisheries in the Northern Atlantic around Iceland have been the site of quite frequent confrontations between the UK and Iceland. (Iceland has long wanted to impose a 200 mile limit around themselves, thereby closing off a large and fish-rich area of the North Atlantic to UK fishermen). In the northeastern USA, there is upset and conflict between people who make their living fishing, and state governments who are trying to preserve fish stocks by limiting the allowed catch. China and Japan are at this moment disputing fisheries in their region. Forests worldwide are rapidly losing territory to the pressure of slash and burn farming. And it goes on. The pressures to provide food, water for drinking and irrigation, power and fuel, are all mounting and there is the feeling of approaching crisis in all those areas. Yet there are no credible plans to solve any of these problems, even though it is quite easy to extrapolate a few years and see where they are all headed. It is also easy (and maddening) to see how our politicians put off the decisions, hoping of course that the problems will not achieve critical mass and explode during their terms of office.

Under stress as well is waste disposal. All the large cities have a continuing problem with where to put the mountains of trash that are produced, and so do many medium and small ones. Local landfills are exceeding their capacity, and trash is having to be hauled at considerable expense, to other places or even other countries. The story about the monster island of trash in the Pacific Ocean was confirmed by a team from CNN in February 2010. (Google "Trash Island Size of Texas". I am not kidding!). Perhaps the most telling and disturbing comment in the CNN report is this:

"It would have been some consolation to think that the Garbage Patch was simply the result of careless sailors, but researchers estimate up to 80 percent of the trash originates on land."

Also mentioned in the article is that this accumulation of trash and other flotsam has always gone on in the Pacific, but that before the advent of so much plastic in our trash, the stuff that accumulated was almost all biodegradable and found its way to the bottom of the ocean and into the food chain from there. The trouble of course is that plastics effectively do not biodegrade, and that is a double tragedy since we desperately need to be recycling our plastics anyway, as before long there will be no oil to make them from, and we will have thrown away so much of what we could still be using.

Much of the environmental stress and the increasing shortages of so many vital resources, are due to the inexorable rise in our world's population. More people means:

- The need for more arable land for food
- More pressure on the world's fisheries
- The need for more drinking water
- The production of more trash

- The need for more fuel
- The need for more electricity
- The need for more housing

The current paradigm actually requires increasing population because it requires more consumption. This, coupled with religious beliefs and cultural traditions that encourage large families, has resulted in tremendous reluctance to control population growth. It is of course a highly sensitive matter to tell people to limit their family size, no matter what their beliefs. Again, the only acceptable way to shift this way of thinking is to shift the worldview.

The Chinese have tried to limit family size by imposing a law to that effect, which has had limited success, except that their culture values boys over girls, so that those couples who have children have been literally aborting girls in favor of boys, resulting in a gender imbalance which will lead to other difficulties as the affected generations grow up.

These are all quite serious problems on their own, and when put together, pose an overwhelming dilemma for us all. It is eloquently described and the obvious implications are put together brilliantly, in Ellen LaConte's book "Life Rules". The subtitle is well worded too. "Why so much is going wrong everywhere at once and how Life teaches us how to fix it".

Let's reiterate things that are going wrong at once, that were listed in Chapter 1. This is probably not a complete list!

- pollution of our water supply

- depletion of the topsoil where crops are intensively farmed

- exhaustion of our water supply (rivers all over the world are no longer reaching the sea for part of the year)

- creation and expansion of deserts and dust bowls
- creation of "dead zones" in the oceans at the mouths of many rivers from fertilizer runoff
- the clearing of huge areas of forest because of the need for more arable land
- the wiping out of fishing grounds and entire fish species from over-fishing
- destruction of important habitat and other natural resources by drilling for oil in such places as the Amazon rainforest, the Niger delta, and Northern Alaska. Not to mention the Gulf of Mexico! Many of the countries of the Middle East have already suffered widespread devastation while their rulers (not their people) have become wealthy.
- Ruination of large areas of places like West Virginia with techniques such as "mountain top removal" to get coal from near the surface. The tar sands of Canada are another case in point.
- Pollution of aquifers with natural gas by breaking up gas-bearing strata to release the gas, which then finds its way into the aquifers and hence to residential drinking water. This results in the awful phenomenon of people being able to light up their household faucets.

Ellen LaConte's conclusion is that we need to "Localize, Localize, Localize". Meaning that we need to coalesce into relatively small communities and to as far as possible, generate our power locally (using renewable and non-polluting technology), grow our food locally, retain and recycle all (or almost all) of our trash, and drastically cut back on all fossil fueled heating, transportation and travel. She also quite rightly advocates minimal importation of manufactured goods or raw materials, doing so only when essential and then doing so by trading with nearby communities. A central part of her proposed scheme of things is the localization, also, of money as far as possible, not necessarily to displace the international system of commerce, but to bring local commerce under local control.

I think we need to place ourselves in a much wider context than our planet. That is, we need to think of ourselves as part of our Galaxy and perhaps even as part of the Universe.

I would like to qualify and expand Ellen LaConte's vision, and this falls squarely into the context of this book. I think we need to place ourselves in a much wider context than our planet. That is, we need to think of ourselves as part of our Galaxy and perhaps even as part of the Universe. Extreme localization is perhaps inevitable if we restrict our thinking and planning to our world only. But it is interesting and I would also say encouraging to note that the ETs that are visiting us are not doing that. They have developed technologies that allow them to travel freely within and probably beyond our galaxy (and by the way, it is abundantly clear that these technologies do not involve fossil fuels or any form of rocketry, nor are they limited by the speed of light). This reveals to us that the access to whatever form(s) of energy that they have, is a major game-changer in terms of making it possible not only to travel but to trade and to get hold of raw materials from asteroids or uninhabited planets without any effective limit.

While that sounds wonderful and is almost certainly achievable for us, it is not without its dangers. It is clear from Ellen LaConte's "Life Rules" that it is not enough to solve one or even several of the problems that humans presently face, unless we solve them all. The point being that on examination, they are all tightly interconnected. So if we get unlimited energy for instance, that would actually result in more rapid population growth and resource depletion. As she so vividly points out, our problems all arise ultimately from a world economy that is based on continual growth, consumption and greed, and all the resource depletion, violence and pollution that go with it. So an overall, one might say holistic, solution is called for.

This makes it even more important that we seek advice from those who have already been in our situation. It is in fact easy to see that every emerging intelligent race will face all of these problems in one form or another and will have to overcome them or endure either extinction or a racial reset in which a combination of war and disease will *at best* set them back thousands of years before they can advance enough to try again. Is our pride so powerful that we would rather risk that than ask for help? Do we want to expose our grandchildren and their descendents to that risk? Because when it gets to their lifetimes, it will be too late.

The same considerations apply to the idea of a middle ground, where we would try to move to a sustainable future without a drastic change in worldview. This idea is pushed by quite a number of people, including of course politicians and others who benefit from the status quo. It is especially seductive to believe that technology can somehow save us. Continuing to use coal but making its use more palatable by extracting the carbon dioxide from the exhaust gases, making it liquid and then storing it underground, is one idea that was described and promoted in the November 2010 issue of Atlantic Monthly. One could of course also do the same kind of thing with natural gas or oil-fired power plants.

The trouble with that approach is twofold. First, it does nothing about the other problems that face us, nor does it do anything about the environmental damage and pollution that result from the mining and extraction processes, to say nothing of the regular spills and explosions that inevitably occur during extraction and transportation (the Exxon Valdez, the BP/Gulf of Mexico disaster, for instance). Second, it takes about 30% of the power output to extract, compress and sequester the carbon dioxide, so that 30% more of the coal, oil, or natural gas is needed to generate the same amount of power.

Similar issues surround the other alternatives to fossil fuels, unfortunately. Solar Photovoltaic power suffers from expensive materials, energy-intensive manufacturing, and the intermittent nature of sunlight. Wind power is also somewhat intermittent, involves expensive manufacturing, installation and maintenance costs, and is restricted in terms of where the resource is available and the cost of getting the power to where it will be used. Nuclear power is hazardous, takes many years to get built and operational, and has an essentially unsolvable disposal problem because the waste takes literally millennia to decay to safe levels. Hydro power is also restricted in terms of location, and requires the taking of large areas of land behind the necessary dam. Tidal power is even more restricted as to location and even then is extremely inefficient because of the low pressure differential driving the turbines. It has also been noted that our infrastructure in the USA is woefully inadequate in terms of distributing any of the following:

- wind power from Texas, the Midwest, or offshore sites
- solar power from the South and Southwest
- tidal power from the Bay of Fundy in Maine
- geothermal power from Yellowstone

Plus, the cost of putting in all those power lines and switching stations has to be reckoned into the cost of that power.

That brings us to the more exotic power sources that are tantalizingly behind the curtain of black budget secrecy. Various individuals worldwide have claimed to have made prototypes of so-called "Free

Energy" devices, and information is available describing the devices and theory behind them both in published books and on the Internet. There are magazines and yearly conferences, but a marketable product has yet to become available.

It is to my mind unfortunate that the term "free energy" is used, because nothing is really free and the term invites ridicule and dismissal. It is also unfortunate that terms like "over unity", "energy from the vacuum" and "perpetual motion" are used, and that statements are made about disproving the law of conservation of energy and the second law of Thermodynamics. If a device is putting out more energy than it apparently takes to keep it in motion, clearly something else is going on! The laws of physics and thermodynamics that are in question apply to closed systems, after all, and if a device is tapping into an energy source from elsewhere, we are no longer talking about a closed system. A useful analogy here is that of the waterwheel at an old mill. The miller is taking energy from outside the mill, and is not putting any energy in, but we can see at once that the energy is coming from outside his system, in this case from the mill stream. Depending on whether the waterwheel is an undershot or overshot type, the stream gives up kinetic or potential energy to the mill, and very simple math will tell us how it all works. It is, in other words, an open system, not a closed one, and that is how we should be viewing these exotic devices. Maybe exotic is too strong a word, too! Let's just say that they are as yet imperfectly understood.

The ability to extract energy from outside of our 3-dimensional world is of course still a remarkable thing, but instead of using language that invites ridicule, we should simply be investigating as busily as we can, and understanding that energy can still be conserved even if we do not (yet) understand where it is coming from. What is abundantly clear from the available information is that there are real effects going on here, and that there are huge potential benefits to be had from studying them.

Most of the talk in this area is about Cold Fusion and Zero Point Energy. Both of these have probably tremendous promise, but to what extent are they really "free"? We will not know until we not only have prototypes but viable manufacturing processes and some experience with operating real machinery over some time. Again, it would be really great to have some advice and guidance from beings that have "been there and done that", since in spite of all kinds of people routinely announcing that "disclosure (of the black ops technologies) is just around the corner", we are likely to still be waiting 50 years from now, largely because the people that are in control of our economic system and the military/industrial complex have no incentive to implement disclosure. It will result in the loss of their enormous power and incomes! We must remember that time is of the essence – it will take a few decades to develop and deploy the new technologies even if government, industry and academia made a full commitment tomorrow. The likelihood of that is low, to say the least. Look at the paltry amount of money that has been invested in the past 40 years by government in solar and wind technologies. For that entire time and even today it has been and is far less than that invested in the scarce, polluting and inefficient coal, oil and natural gas technologies.

It could be and probably will be argued that hoping for technical help from friendly ET races is another form of believing that technology will save us. But that must not be all that we ask for, just because our various problems are indeed inter-related. What we need from our neighbors is much more than technology. We desperately need a knowledge and understanding of how they (and the other races that made it) got through their version of this crisis.

What, exactly, do we have to lose by seeking that help? We certainly have a hell of a lot to lose by not seeking it!

Anyway, it should go without saying that while we seek that help, we still have to keep searching, experimenting, exploring and learning as much as we can as fast as we can, in case the help we are hoping for is not going to be forthcoming for whatever reason.

Chapter 8

What can we look forward to, longer term?

"A Stand can be made against Invasion by an Army; no Stand can be made against Invasion by an Idea." Victor Hugo

Abstract: The new technologies and solutions to our various problems will come to us in time, even if a relatively small number of people take part in the dialogue with the ETs. In other words, the good news is that there is no way that the dialogue can be suppressed. The medium term will probably be one of sometimes painful adaptation to the changes in technology and ideas that will be coming rapidly at us. We will start interstellar travel and interstellar commerce. There will be fundamental change in the way almost all of us relate to each other and to the Universe. There will inevitably be holdouts, (even today there are people who still believe the Earth is flat), but the evidence and the benefits will be pervasive and overwhelming. Most importantly, it will be easily possible for everyone to directly experience those benefits if they wish.

This closing chapter is necessarily going to be more speculative than the preceding ones, but I think the reader will find that there is plenty of information to be found about the longer term prospects for our race. A goodly number of these can be found in the Bibliography section at the end, and a few of them are mentioned below.

Before going any further I want to say a little about reincarnation. At first glance that would seem to have little to do with ETs and our interactions with them, but please bear with me! You see, if we are destined to return here for future lives, the whole thing becomes a rather different game. The basic issues are still essentially the same –

we want to leave this world a better place for our grandchildren and their descendants, for sure. However there is a considerable change in perspective when we realize that we ourselves will be coming back. It becomes a whole lot more important that we take a very long view of how we treat our planet and each other, and of course how we view our role in the galaxy and the Universe, which is where our relationship with our ET neighbors comes in. It means that it makes sense in a whole new way to make our ongoing home a better place.

A great many mystics, philosophers and other well-known people, including arguably Jesus of Nazareth, (depending on how you interpret some of his sayings), have argued for the idea that we have many lives. A whole sect of the early Christian church, the Gnostics, believed this and were persecuted for it. The early church had decided that they would promote the idea of a single life on which everything depends. The church leaders of that time realized that they stood to lose a great deal of their power if their followers understood that they were responsible for their own lives. With everything hanging on ending this single life in a state of grace, (that state of grace as defined by the priesthood, naturally) the priesthood holds the power because they are the ones setting the rules for ascension – interesting, and hardly surprising when you think about it!

Moving now to more recent history, reincarnation is a central tenet of the Hindu and Jainist religions. It was believed in firmly by such diverse people as Henry Ford, General George Patton, and Saddam Hussein (!). The Dalai Lama is himself an ongoing reincarnation that goes back many hundreds of years. Other historical figures of note whose sayings and writings indicate the same belief, are Cicero, William Shakespeare, Giordano Bruno, Johann Goethe, Arthur Schopenhauer, William Butler Yeats, William James, Henry David Thoreau, Ralph Waldo Emerson, Walt Whitman, Thomas Huxley, Carl Jung, and Rudolf Steiner. Modern proponents include Edgar Cayce,

Jane Roberts (channeling Seth), Carla Rueckert (channeling Ra, see her site www.llresearch.org), L. Ron Hubbard (Scientology), David Wilcock (see his site www.divinecosmos.com), Dr. Brian Weiss who has published several books on past lives and lectures regularly on his work (see his site www.brianweiss.com), Shirley MacLaine (www.shirleymaclaine.com), where she states that three-quarters of the world's population believes in reincarnation), the well-known authoress Taylor Caldwell, the Eckankar group (www.eckankar.org), and the Theosophical Society. By the way I am neither promoting nor endorsing any of these people or organizations, except that I have personally heard Carla Rueckert, Brian Weiss, and David Wilcock, and was impressed by all of them. There are many practitioners who will get you information on past lives using hypnotic regression. Not all of them are legitimate, of course, any more than are, say, all lawyers, doctors, authors, engineers, investment advisors, or politicians! One has to take care in selecting a teacher or counselor just the same as any other professional.

There are several reasons that I find the idea of reincarnation very persuasive. Firstly, it handles clearly and elegantly the problems posed by infant deaths where the child has had no opportunity to choose a belief system or a life path, and the related issue of those people who because of the place and/or culture in which they live, also live their lives in ignorance of the choices that apparently need to be made. Secondly, it deals with the rather absurd notion that we can learn everything we need to know in a single lifetime. Thirdly, by presenting us with a series of lives in which we are born to various races, cultures and environments, both genders, riches and poverty, and various skills and talents, reincarnation gives us a complete education and the chance to learn all the important lessons. Fourthly, it gives us full responsibility for the consequences of our actions, and the opportunity to take as many or as few lives as we need to grow and develop into beings that are ready to advance to the next stages of growth and development, whatever they might be.

It is thus a richer and vastly more equitable approach to life. It brings a much more profound meaning to the Bible quote, "As ye sow, so shall ye reap" (Jeremiah 1-19).

The above is a very brief discussion. If you look around and do some reading and study, you will find plenty of reason to accept that all of us are in a growth and learning process involving many, many lives, which means that we are going to have to keep working on taking care of our planet as well as our own personal growth for a very long time, effectively forever.

If we continue on the path of the fast buck, ever increasing consumption, ever growing population, and continuous inflation, and we use up and eventually fight over our dwindling resources, there is an inevitable result, and it is truly amazing that almost everyone is unwilling to see it coming. It is a mass denial of the obvious that staggers the imagination.

So I will spell it out. We will end up losing our civilization, setting us back thousands if not tens of thousands, of years. Depending on how reincarnation works, that probably means that we get to live probably hundreds more lives through that long climb back to another chance at getting it right. If we believe the story of Atlantis, we have been through that cycle at least once before, perhaps several times. How about making it to the next level this time? The Atlantis collapse, by the way, according to many authors including the late Zecharia Sitchin, did not involve a complete reversion to primitive living, in that there is evidence that a relatively small number of the Atlanteans survived and their descendants have passed down information about their capabilities and technologies to this day. That does not seem to have done the rest of us much good, however. It seems that the Atlantean survivors and their descendants decided that power and wealth were preferable to the satisfaction of helping their fellow men and women.

The people who study this are somewhat divided on whether the move up to the next level, sometimes known as fourth density (first density is inanimate objects such as rocks, second density is the animal kingdom, and third density is where we are now), will occur as a group event when some critical mass of people have arrived at the required level of development, or whether there will be a separation between those people who have reached that level and those who have not, with the "graduates" moving on and the rest having to repeat this "grade level". In cosmic terms, we need to accept that this density/grade level is effectively kindergarten, and fourth density is effectively first grade. We have a long way to go! It is reassuring, though, to know that there is always another chance, in one form or another, and in one place or another. Having to repeat kindergarten after all, is a setback for a child, but not the end of his or her world!

Perhaps it is necessary to go through several tries at reaching the next level. If we imagine that all of us are learning and growing with each lifetime (with the occasional step backwards, I suppose) there will be a steady overall improvement. Each time this cusp that we are now on is reached, more people will have learned and grown enough to personally reach the required level. Eventually enough of us will have achieved that and raised the overall level high enough to "graduate". Again, it would be very helpful if we can find out from our ET friends how the process worked for them and others in the galaxy, and how we can help ourselves.

It is possible that the notion of repeated cycles with a step up the evolutionary ladder each time, is a version of General Periodicity as proposed by August Jaccaci in his book with that title. The proposition, briefly, is that every organism from the most primitive (molluscs, butterflies) to the most complex (businesses, political parties, civilizations), goes through a repeated sequence of stages which he calls Gather, Repeat, Share, and Transform. The application of that idea to the development of the human race would place us presently at the Share stage, moving into Transformation. The nature of that transformation is still to be determined!

There is an even worse case in that we could possibly destroy the human race on this planet completely. Even then, we will continue with lives on other planets, but we will have lost this one, which would be a loss of literally galactic proportions. Although there are apparently hundreds of thousands, more likely many millions, of intelligent races in this galaxy, there cannot be very many in our particular stage of development at any one time, and the loss of any one of those must cause immense grief galaxy-wide. The risk of that loss I believe is what makes so many other races in the galaxy willing to help us. But it definitely seems that we have to be willing and humble enough to ask for help, and after we have undergone the training which will enable us to access that help, to listen and to learn!

It may be useful to make a rough calculation to get an idea of how unique our situation at this time may be. If we suppose that there are at any one time 100,000 races between initial sentience and a state of development at which they have evolved beyond existing as individuals, and that the duration of that climb is 500,000 years, that gives us a rough basis on which to work. It looks as if the crisis period lasts a few hundred years, say on average 300 (about the time between the first emergence of technology and the present day). Let us also assume that the crisis occurs 5 times for each race between 10,000 and 100,000 years into the process. So 1500 years out of 90,000 is 1 in 60. In that case we could say that 60 or so races are in crisis at any one time. That is a very few over a whole galaxy. Lots of assumptions, I know, but I do find it useful to put the scheme of things into some kind of context.

So yes, the information, the technology, and the move up to the next level, will indeed come in time, and that is the good news. The bad news is that it is liable to come with some pain and suffering, and that it may not be this time around for many of us, if we blow this chance and have to repeat the cycle. It is likely that again some number of people will have planned ahead or will be just lucky enough to survive with enough in the way of technology and tools, to get a good start on the long climb back. Let us hope that they will be more willing to share their knowledge than the surviving Atlanteans were.

Perhaps it will encourage some of us to put more energy into this process if we imagine how our world will look if we can take the high road. The new technology and awareness will be amazing and extremely rewarding. We will start interstellar travel and interstellar commerce. There will be fundamental change in the way almost all of us relate to each other and to the Universe, in that it will be so much easier to communicate clearly and honestly if we do so mind-to-mind. This more powerful communication will enable much more rapid and comprehensive education in as many subjects as we desire. It will be possible to travel widely, not only on our world but around the galaxy and probably beyond. We will meet and interact with not only those races that have assisted our evolution and been our mentors, but also with races that are still developing and that we can help in turn.

All this may sound very airy-fairy and idealistic. Perhaps it is. But perhaps it is not! We risk nothing by finding out from our ET neighbors how it all works and has worked for them and others. It is time to reach out, carefully but determinedly, for our destiny!

Bibliography

Backster, Cleve: *Primary Perception.* White Rose Millennium Press (2003)

Bernstein, Paul: *Intuition: What Science Says (so far) about how and why Intuition Works.* Endophysics, Time and the Subjective. World Scientific (2005)

Brown, Dr. Courtney: *Cosmic Voyage.* Penguin Onyx (1997)

Brown, Dr. Courtney: *Remote Viewing.* Farsight Press (2005)

Buchanan, Lyn: *The Seventh Sense.* Pocket Press (2003)

Campbell, Colin: *Oil Crisis.* Multi-Science Publishing (2005)

Corso, Philip J., US Army Colonel (Ret.): *The Day After Roswell.* Pocket Books, Simon & Schuster (1997)

Friedman, Stanton: *Flying Saucers and Science.* New Page Books (2008)

Friedman, Stanton, and Berliner, Don: *Crash at Corona: The US Military Retrieval and Cover-Up of a UFO.* Paraview (2004)

Gore, Al: *An Inconvenient Truth.* Rodale Books (2006)

Greer, Dr. Steven M.: *Disclosure.* Crossing Point Publications (2001)

Greer, Dr. Steven M.: *Hidden Truth – Forbidden Knowledge.* Crossing Point Publications (2006)

Greer, Dr. Steven M.: *Contact : Countdown to Transformation.* Crossing Point Publications (2009)

Grimsley, Ed: *IUFOC 2008 Presentation*

Grinspoon, David: *Lonely Planets.* Harper Perennial (2004)

Haines, Richard F.: *CE-5.Close Encounters of the Fifth Kind.* Sourcebooks (1999)

Hopkins, Budd: *Sight Unseen.* Pocket Star (2004)

Jaccaci, August: *General Periodicity.* Unity Scholars Media (2000)

Kean, Leslie: *2006 Chicago O'Hare UFO.* The Huffington Post (1/4/2011)

Krapf, Phillip H.: *The Challenge of Contact.* Origin Press (2003)

Larkins, Lisette: *Listening to Extraterrestrials.* Hampton Roads Publishing (2004)
Larkins, Lisette: *Calling On Extraterrestrials.* Hampton Roads Publishing (2003)
LaConte, Ellen: *Life Rules.* iUniverse (2010)
LoBuono, George: *Alien Mind.* QC Press (2010)
Mack, John E.: *Passport to the Cosmos.* Three Rivers Press (1999)
Mannion, Michael: *Project Mindshift.* M. Evans and Company (1998)
McMoneagle, Joe: *Remote Viewing Secrets.* Hampton Roads Publishing (2000)
Mitchell, Edgar F.: *Nature's Mind - The Quantum Hologram.* Article published by the National Institute for Discovery Science (2003)
Nelson, Michael: *Portage, Ohio UFO Chase.* MUFON Denver Presentation (2007)
Radin, Dr. Dean: *The Conscious Universe.* HarperOne (2009)
Radin, Dr. Dean: *Entangled Minds.* Pocket Books, Simon & Schuster (2006)
Reich, Wilhelm: *Selected Writings: An Introduction to Orgonomy.* Farrar, Straus & Giroux (1963)
Rhine, Dr. J.B.: *Reach of the Mind.* William Morrow & Co (1979)
Roberts, Jane: *The Seth Material.* New Awareness Network (2010)
Rueckert, Carla, Elkins, Don, and McCarty, Jim: *The Ra Material.* Donning Company Publishers (1984)
Sagan, Carl: *Contact.* Simon & Schuster (1985)
Sitchin, Zecharia. *The Earth Chronicles Series.* Harper (2007)
Sitchin, Zecharia. *The End of Days.* HarperCollins (2007)
Sitchin, Zecharia. *The Case of Adam's Alien Genes.* Article on the Internet at www.sitchin.com/adam.htm
Sparks, Jim: *The Keepers.* Wild Flower Press (2006)
Strieber, Whitley: *Communion.* Harper Paperbacks (2008)
Strieber, Whitley: *Confirmation: The Hard Evidence of Aliens Among Us.* St Martin's Press (1998)
Targ, Russell and Puthoff, Harold: *Mind-Reach.* Hampton Roads Publishing (2005)

Tesla, Nicola and Childress, David Hatcher: *The Tesla Papers.* Adventures Unlimited Press (2000)
Von Daniken, Erik: *Chariots of the Gods.* Berkley Trade (1999)
Von Ward, Paul: *Gods, Genes and Consciousness.* Hampton Roads Publishing (2004)
Webre, Alfred: *Exopolitics.* UniverseBooks (2005)
Weiss, Dr. Brian: *Many Lives, Many Masters.* Fireside (1988)
Wilcock, David: *The Science of Peace.* On line at www.divinecosmos.com (2007)

Resources

Many resources are available on the Internet. Google and other search engines will yield useful if overly numerous results to simple queries. Wikipedia has lots of information, definitions and origins of names and ideas such as Close Encounters, Reincarnation, and Remote Viewing. It is more useful if you want to get information about a tightly targeted subject in that it will give you a single response rather than the sometimes hundreds of thousands of items that the search engines will come up with. Amazon.com, Alibris.com and others will locate both in print and out of print books very easily. Some more specific web sites that are relevant to this book are listed below. They are all active as this book goes to publication.

www.brianweiss.com
- has information and discussion on his ongoing work, books, lectures, appearances. etc.

www.cropcircleconnector.com
- has up-to-date pictures, information and discussion on crop circles worldwide.

www.cseti.org
- is the site for Dr. Steven Greer and his ongoing work. Information on Ambassador to the Universe trainings, photos and videos of ET contact, and a number of publications.

www.divinecosmos.com
- is the site for David Wilcock and his ongoing work, with many free downloads, as well as publications

for sale, information on appearances and seminars, and regular commentary on world events.

www.earthpolicy.org
- is the web site of the Earth Policy Institute, which publishes Lester Brown's Plan B 4.0. His Plan B books are updated and re-issued regularly as events relating to our planet's health unfold. Earth Policy also puts out regular publications on environmental issues.

www.farsight.org
- is the site for Courtney Brown's work in Remote Viewing. Training material and publications, some with free download.

www.llresearch.org
- is the site for Carla Rueckert's ongoing work in channeling ETs and other beings, and is where the Ra material and other publications may be found.

www.mindshiftinstitute.org
- carries articles and information on a lot of topics related to this book, including material from people who have experienced ET contact.

www.sitchin.com
- is the site for Zecharia Sitchin's work. Although sadly Mr. Sitchin passed away late in 2010, it still has a great deal of information and makes available his books and lectures.

And lastly –

www.extraterrestrials-thebook.com
- the site for this book. Information including upcoming events and appearances, plus a book excerpt and synopsis. Regularly updated reporting on recent events of interest.

ABOUT THE AUTHOR

Stephen Mather-Lees, MA, MIEEE was educated in England at Caterham School and St. Catharine's College, Cambridge. He is married with two grown children and lives in Southern NH.

He has always been at the cutting edge of technology, and his degree in Mechanical Sciences (General Engineering) has enabled him to rapidly become expert in many technologies and disciplines during his career. Apart from a deep and abiding interest in who we are, where we came from, and where we are going, he is involved in environmental issues and the technologies that relate to alternative energy.

Made in the USA
Charleston, SC
13 November 2011